Hydrological Models for Environmental Management

NATO Science Series

A Series presenting the results of activities sponsored by the NATO Science Committee. The Series is published by IOS Press and Kluwer Academic Publishers, in conjunction with the NATO Scientific Affairs Division.

A. Life Sciences	IOS Press
B. Physics	Kluwer Academic Publishers
C. Mathematical and Physical Sciences	Kluwer Academic Publishers
D. Behavioural and Social Sciences	Kluwer Academic Publishers
E. Applied Sciences	Kluwer Academic Publishers
F. Computer and Systems Sciences	IOS Press

As a consequence of the restructuring of the NATO Science Programme in 1999, the NATO Science Series has been re-organized and new volumes will be incorporated into the following revised sub-series structure:

I. Life and Behavioural Sciences	IOS Press
II. Mathematics, Physics and Chemistry	Kluwer Academic Publishers
III. Computer and Systems Science	IOS Press
IV. Earth and Environmental Sciences	Kluwer Academic Publishers
V. Science and Technology Policy	IOS Press

NATO-PCO-DATA BASE

The NATO Science Series continues the series of books published formerly in the NATO ASI Series. An electronic index to the NATO ASI Series provides full bibliographical references (with keywords and/or abstracts) to more than 50000 contributions from international scientists published in all sections of the NATO ASI Series.
Access to the NATO-PCO-DATA BASE is possible via CD-ROM "NATO-PCO-DATA BASE" with user-friendly retrieval software in English, French and German (WTV GmbH and DATAWARE Technologies Inc. 1989).

The CD-ROM of the NATO ASI Series can be ordered from: PCO, Overijse, Belgium

Series 2. Environmental Security – Vol. 79

Hydrological Models for Environmental Management

edited by

Mikhail V. Bolgov
Water Problems Institute,
Moscow, Russia

Lars Gottschalk
Department of Geophysics,
University of Oslo, Norway

Irina Krasovskaia
IGC, Sweden

and

Robert J. Moore
Centre for Ecology and Hydrology,
Wallingford, U.K.

Springer-Science+Business Media, B.V.

Proceedings of the NATO Advanced Research Workshop on
Hydrological Stochastic Models of Processes and their Applications in Problems of
Environmental Preservation
Moscow, Russia
23–27 November 1998

A C.I.P. Catalogue record for this book is available from the Library of Congress.

ISBN 978-1-4020-0911-2   ISBN 978-94-010-0470-1 (eBook)
DOI 10.1007/978-94-010-0470-1

Printed on acid-free paper

CONTENTS

Preface

This book contains a selection of papers from a NATO Advanced Research Workshop entitled "Stochastic models of hydrological processes and their applications to problems of environmental preservation" convened in Moscow over the period 23-27 November 1998. The Workshop was unique in providing the first opportunity for over a decade for countries of the Russian Federation to interact with other countries across the world to discuss hydrological science issues relevant to environmental management. The contrasting schools of thought within the Russian Federation and with other countries proved a fascinating and valuable experience for those fortunate enough to attend.

The scientific content of the Workshop was motivated by a number of concerns. Water is a key natural resource whose modelling and management is made complex by its inherent spatial unevenness and time variability. Traditional methods for investigating hydrological processes in nature employ stochastic modelling and forecasting. However these are not well developed with regard to (i) representing the characteristics of hydrological regimes, and (ii) investigating the influence of water factors on processes which arise in biological systems and those involving hydrochemical, geophysical and other processes.

Advances in modelling and risk assessment are required to develop technical standards for permissible loadings on our environment. In particular, existing methods are deficient for forecasting the regime of ecological systems involving water resources affected significantly by man-made impacts. With environmental preservation of growing concern in many countries, the NATO Advanced Research Workshop provided a timely opportunity to address the modelling, forecasting and risk assessment issues that this management problem presents from a multi-disciplinary viewpoint.

The workshop provoked discussion on ideas for new stochastic modelling approaches against a requirement for sustainable development in an environment changing under man's influence. A particularly important challenge was seen to be the need to consider the effects of a changing climate and ecological impacts when developing modelling and risk assessment procedures in support of river basin management.

The book "Hydrological models for environmental management" contains 18 papers chosen to cover a good cross-section of the science issues addressed by the NATO Advanced Research Workshop.

Mikhail V. Bolgov, Water Problems Institute, Moscow, Russia
Lars Gottschalk, Department of Geophysics, University of Oslo, Norway
Irina Krasovskaia, IGC, Sweden
Robert J. Moore, Centre for Ecology & Hydrology, Wallingford, UK

Acknowledgements

This NATO Advanced Research Workshop was directed by Mikhail J. Bolgov, Water Problems Institute, Moscow, Russia and Robert J. Moore, CEH, Wallingford, UK with the scientific support of Lars Gottschalk, Department of Geophysics, University of Oslo, Norway as the third member of the Organising Committee. Irina Krasovskaia, IGC, Sweden provided scientific and editorial support relating to manuscripts from the Russian Federation. The Workshop was hosted by the Water Problems Institute and convened at a hotel of the Russian Academy of Sciences.

The NATO Science Committee is thanked for their support of the Workshop. NATO gave grant support with employers of the members of the Organising Committee bringing additional resources. Participants in the Workshop, 34 from eight countries of the former Soviet Union and 10 from nine countries across the world, are particularly thanked for their contribution to its success.

List of Participants

Directors

Robert J. Moore
Centre for Ecology and Hydrology
Wallingford
Oxon, OX10 8BB
UK

Mikhail V. Bolgov
Water Problems Institute
117971, Gubkin St., 3, Moscow
Russia

Key Speakers

Vit Klemes
3460 Fulton Rd, Victoria
Canada

Lars Gottschalk
Department of Geophysics
University of Oslo
P.O.Box 1022 Blindern
N-0315 Oslo
Norway

Daniel Loucks
Hollister Hall
Cornell University
Ithaca, NY 14853
USA

Ladislav Kashparek
T.G.M. Water Research Institute
Podbabska 30,160 62 Prague 6
Czech Republic

Atil Bulu
Istanbul Technical University
Civil Engineering Faculty
Hydraulics Division
80626 Ayazaga, Istanbul
Turkey

Irina Krasovskaia
IGC
Lergravsvagen 33
S-26352 Hoganas
Sweden

Daniil Ratkovich
Water Problems Institute
117971, Gubkin St., 3, Moscow
Russia

Josef Khranovich

Water Problems Institute
117971, Gubkin St.,3, Moscow
Russia

Anatoli Rozhdestvensky

State Hydrological Institute (GGI)
199053, St. Petersburg
Vasilievski ostrov, 2-nd line St., 23
Russia

Yurii Vinogradov

State Hydrological Institute (GGI)
199053, St. Petersburg
Vasilievski ostrov, 2-nd line St., 23
Russia

Vladimir Debolskii

Water Problems Institute
117971, Gubkin St.3, Moscow
Russia

Sergey Kondratyev

Institute of Limnology
Russian Academy of Sciences
196199, St. Petersburg
Sevostyanova St., 9
Russia

Evgenii Venezianov

Water Problems Institute
117971, Gubkin St.3, Moscow
Russia

Vladilen Pisarenko

International Institute of Earthquake
Theory and Prediction
Russian Academy of Sciences
113556, Warshavskoe Sh., 79, korp.2
Moscow, Russia

Nikolai Alekseevskii

Moscow State University
Geography Department
119899, Vorob'evy Gory
Moscow, Russia

Val. Pryazhinskaya

Water Problems Institute
117971, Gubkin St. 3, Moscow
Russia

Guozhang Feng

College of Water Resources and Architectural
Engineering
Northwestern Agricultural University
Yangling, Shaanxi, 712100
China

Nikolay Bulgakov

Moscow State University
Biological Department
119899, Vorob'evy Gory
Moscow, Russia

Other Participants
Natalia Loboda

Odessa Hydrometeorological Institute
270016, Odessa –16
Lvovskaya St., 15
Ukraine

Leonid Koritnii

Institute of Geography
664033 Irkutsk
Ulanbatorskaya St., 1
Russia

Yurii Trapeznikov

Institute of Limnology
Russian Academy of Sciences
196199, St. Petersburg
Sevostyanova St., 9
Russia

Yaroslav Ivanio

Agricultural Institute
664038, Irkutsk,
pos.Molodezhnii, 2, ap.9.
Russia

Pieter Van Gelder

Delft University of Technology
Faculty of Civil Engineering
Stevinweg 1, Room 3.87
P.O. Box 5048
2600 GA Delft
The Netherlands

Pavel Petrovich

VUVH
N.g. Svobodu 5, 81249
Bratislava
Slovakia

Boris Gartsman

Pacific Ocean Institute of Geography
690041, Vladivostok
Radio St., 7.
Russia

Hafzullah Aksoy

Istanbul Technical University,
Civil Engineering Faculty
Hydraulics Division
80626 Ayazaga, Istanbul
Turkey

Knarik Oganesyan

Water Problems Institute
375047, Erevan
Amaranocain – 125
Armenia

V. Rumyantsev

Institute of Limnology
Russian Academy of Sciences
196199, St. Petersburg
Sevostyanova St., 9
Russia

Tatyana Vinogradova

State Hydrological Institute (GGI)
199053, St. Petersburg
Vasilievski ostrov, 2-nd line St., 23
Russia

Nikolay Vakhonin

Institute for Melioration
Gor'kogo St., 153
Minsk
Belorus

Alexandre Pak

SANIGMI
700052, Tashkent
72, K. Maksumov St., 72
Uzbekistan

Evgenii Evstigneev

Moscow State University
Geography Department
119899, Vorob'evy Gory
Moscow, Russia

Natalia Penkova

State Hydrological Institute (GGI)
199053, St. Petersburg
Vasilievski ostrov, 2-nd line St., 23
Russia

Andrey Khristoforov

Moscow State University
Geography Department
119899, Vorob'evy Gory
Moscow, Russia

G.V. Matushevskii

State Institute of Oceanography
119838, Moscow
Kropotkinsky per., 6
Russia

Farda Imanov

Baku University
Hydrometeorological Department
370145, Baku-145, Z.Khalilova St., 23
Azerbaijan

Gasan Fatullaev

Institute of Geography
370143, Baku, G.Javida St., 31
Azerbaijan

Elena Asabina

ROSNIIVH
620049, Ekaterinburg
Mira St.,23
Russia

A.I. Gavrishin

Novocherkassk Technology University
Department of Hydrogeology
346400, Novocherkassk
Prosvrsheniya St., 132
Russia

Luc Perreault

ENGREF
19 Avenue du Main
75732 Paris CEDEX 15
France

Nikolay Filatov

North Water Problems Institute
Karelia Scientific Centre
185 030, Petrozavodsk
Urickogo St.,50
Russia

Oleg Savichev

Tomsk Institute of Oil and Gas
635 055,Tomsk, Box. 2282
Russia

Anatoly Frolov

Water Problems Institute
117971, Gubkin St., 3, Moscow
Russia

Vladimir Nikora

NIWA
10 Kyle Street, PO Box 8602
Christchurch
New Zealand

BAYESIAN ANALYSIS OF A SIMULTANEOUS CHANGE IN BOTH THE MEAN AND THE VARIANCE

Luc PERREAULT[a,b], Eric PARENT[a]

[a]GRESE Laboratory

ENGREF, 19 Avenue du Maine

75732, Paris, France

Bernard BOBEE[b]

[b]Chaire en hydrologie statistique Hydro-Québec/CRSNG

INRS-Eau, 2800 rue Einstein CP 7500

Ste-Foy, Canada G1V 4C7

1   Introduction

Although such an hypothesis is rarely stated explicitly, the assumption that stochastic time series are stationary plays a crucial role in water resources management. Under the assumption that tomorrow will statistically behave like yesterday, stochastic models are fitted to hydrometeorological variables such as river flow, precipitation and temperature. The estimated models are then used for many engineering purposes, in particular for simulating the operation of hydropower systems (energy planning, design of power plants, operation of reservoirs). Consequently such models, and the decisions stemming from them, are based on the assumption of stationary behavior of hydrometeorological inputs. However, many climatologists believe that there is more and more evidence that serious changes have occurred in many parts of the world, which leads to questioning the stationarity hypothesis in hydrometeorological time series analysis.

While the problem of a change in the mean of a sequence occurring at an unknown time point has been addressed extensively in the hydrological literature (among others, Bruneau and Rassam, 1983; Bernier, 1994; Lubès-Niel et al., 1998; Perreault et al. 1999a), the problem of a change in the variance has been less widely covered. In hydrology, to our knowledge, it has not been studied from a Bayesian perspective although the variance of hydrological series is an important parameter in water resources management. In particular, a change in the fluctuations of net basin supply may arise from an increase in the occurrence of extreme events. Analysis of changes in the variance of net basin supplies is important because it will have a direct impact on reservoir operation and capacity sizing.

In this paper, interest is mainly focussed on the estimation of the unknown time-point and intensity of a simultaneous single change in both the mean and

1

M.V. Bolgov et al. (eds.), Hydrological Models for Environmental Management, 1–15.

the variability of a sequence of normal random variables. The aim is to outline that this problem can be straightforwardly addressed through the Bayesian framework, which in turn can be easily solved by implementation of explicit inference technics.

In Section 2, we formulate the general Bayesian change-point setting. To illustrate the principles, we derive in Section 3 the posterior distributions of the parameters of interest using conjugate prior distributions. It is seen that these posterior distributions are expressible in closed form and therefore results to make inference are directly available. In Section 4, the approach is applied to a series of 54 annual energy inflows. Finally, Section 5 offers a general discussion.

2 Bayesian change-point analysis

In this section, the general Bayesian change-point problem is formulated. Some advantages of the Bayesian approach are first recalled.

2.1 WHY BAYESIAN ANALYSIS?

In almost any water resources management problems, decisions have to be taken in situations of uncertainty. The decision maker has to choose an alternative from a set of possible actions (e.g. construct a new power plant or not), and each action involves a range of uncertain consequences with social and economic aspects. To make a choice, the decision maker must quantify uncertainty about the studied phenomenon (existence of a change, intensity of the shift if any, etc.). Since we are also concerned to avoid making "illogical" choices, his degree of belief should be based on some rationale to ensure good representation and communication of uncertainty. In this paper, we assume that the decision maker is a rational person in the sense of the coherence axioms given by Savage (1972). Such a definition of rationality has been discussed by Walley (1991) and Bernardo and Smith (1994) in a general context. Of course, we do not pretend that all individuals face to an uncertainty strictly respect these axioms. Munier and Parent (1998) gave counter examples in hydrology and discussed various deviations from these axioms. Nevertheless, such an axiomatic framework is, at least in first approximation, a good tool for consistent representation and communication of degrees of belief about a given event.

The use of probability is very common to describe observable variables (data). In the previous setting, it can be extended to quantify the degree of belief about unobservable quantities (unknown parameters) by the mean of a probability distribution (prior distribution). Therefore, subjective knowledge or other sources of informal specific information in addition to the data itself, can also be integrated in the hydrological analysis (expertise, regional information, past analysis, etc.). Acceptance of Savage's axiomatic leads us naturally to the Bayesian approach, since its essence is precisely to update such prior beliefs with the information carried out by the data. This is formally done by using the Bayes theorem which combines the prior distribution (degree of belief about the unobservable parameter of interest) and the likelihood function (degree of belief about the observable data) to evaluate the posterior distribution. Updating beliefs looks appealing to most decision makers, since it mimics their own behavior in risky situations.

As it will be demonstrated in the rest of the paper, the Bayesian approach is convenient and quite straightforward for change-point analysis. Some additional technical advantages are listed below.

1. Bayesian analysis yields a final distribution (the posterior distribution) for the unknown parameters, and from this, several questions can be answered simultaneously. For instance, one can not only estimate the parameters, but can directly obtain accuracy measures for the estimate. This is in contrast to classical statistics, for which obtaining estimates and determining the accuracies are two different problems.

2. Interpretation of the results is easy and straightforward in the Bayesian framework compared to classical analysis. The frequentist interpretation is difficult to understand for a non-statistician, and water resources managers often find it too much abstract for operational concerns. For example, interpretation of classical confidence intervals involves averaging over all possible data, while it is known which data occurred. In fact, practitioners usually interpret the results of a classical confidence interval as a Bayesian credible interval (Lecoutre, 1997).

3. Hydrologists are prone to illustrate the performance of their decisions under a variety of scenarios. Bayesian analysis is particularly suited to generate such scenarios by use of the predictive distribution. An application of predictive analysis in the context of change-point analysis for hydrological series can be found in Perreault et al. (1999b).

2.2 GENERAL FORMULATION OF CHANGE-POINT PROBLEMS

The simplest formulation of the change-point problem is the following. Let us assume that the densities p_1 and p_2 belong to a known parametric class of probability densities $\mathbf{P} = \{p(x \mid \theta); \theta \in \Theta\}$ indexed by an unknown parameter θ such that, for a sequence of n independent random variables $\mathbf{X} = (X_1, X_2, ..., X_n)$, we have

$$
\begin{aligned}
X_i \sim p_1(x_i) &= p(x_i \mid \theta_1), i = 1, ..., \tau \\
X_i \sim p_2(x_i) &= p(x_i \mid \theta_2), i = \tau + 1, ..., n
\end{aligned}
\tag{1}
$$

where $\theta_1 \neq \theta_2$ and $\tau = 1, 2, ..., n-1$ is an unknown parameter, called the change-point. That is, the first and second parts of the sequence of random variables are distributed as statistical distributions which belong to the same class, but with different values of the unknown parameter θ. Since the unknown change-point τ can take values between 1 and $n-1$, this model assumes a change occurred with certainty. The likelihood function resulting from n observations $\mathbf{x} = (x_1, x_2, ..., x_n)$ becomes

$$p(\mathbf{x} \mid \theta_1, \theta_2, \tau) = \prod_{i=1}^{\tau} p(x_i \mid \theta_1) \prod_{i=\tau+1}^{n} p(x_i \mid \theta_2) \qquad (2)$$

from which, for instance, maximum likelihood estimates for τ, θ_1 and θ_2 can be obtained (see for example, Hinkley, 1970). In the Bayesian perspective, a joint prior distribution $p(\theta_1, \theta_2, \tau)$ is assumed for the parameters, and using the Bayes theorem it is seen that the joint posterior distribution $p(\theta_1, \theta_2, \tau \mid \mathbf{x})$ of θ_1, θ_2, τ given the data is proportional to

$$p(\mathbf{x} \mid \theta_1, \theta_2, \tau) \times p(\theta_1, \theta_2, \tau) \qquad (3)$$

Just as the prior distribution $p(\theta_1, \theta_2, \tau)$ reflects beliefs about the parameters prior to experimentation, the posterior distribution $p(\theta_1, \theta_2, \tau \mid \mathbf{x})$ reflects the updated beliefs after observing the sample data. In the Bayesian framework, all statistical inference about the unknown parameters is based on the posterior distribution (Berger, 1985).

Interest is now on making inference about the change-point τ, θ_1, θ_2, and any suitable function $W(\theta_1, \theta_2)$ which can describe the amount of shift. To do this, all marginal posterior distributions must be found by integration. For example, to evaluate $p(\tau \mid \mathbf{x})$, θ_1 and θ_2 must be integrated out of $p(\theta_1, \theta_2, \tau \mid \mathbf{x})$. Using conjugate prior distributions, and assuming prior independence between τ and the other parameters (θ_1, θ_2), i.e. $p(\theta_1, \theta_2, \tau) = p(\theta_1, \theta_2) p(\tau)$, solutions in closed form can be obtained for some simple models (it is the case for a single change in both the mean and the variance for the normal model). A prior distribution $p(\theta)$ is said to be a conjugate density for $p(x \mid \theta)$ if the posterior distribution $p(\theta \mid x)$ belongs to the same class of density functions as $p(\theta)$ (Berger, 1985). This can be achieved by choosing $p(\theta)$ to have the same structure as $p(x \mid \theta)$, when the latter is viewed as a function of θ. Likelihoods for which conjugate prior density functions exist are those corresponding to exponential family models (Bernardo and Smith, 1994). In more complex models, even with conjugate prior and independence, integration may be a very difficult analytic task. However, use of Markov Chain Monte Carlo methods such as the Gibbs sampler (Gelfand and Smith, 1990), which is particularly suitable for change-point analysis, enables a straightforward solution to such problems (Carlin et al., 1992; Stephens, 1994; Perreault et al. 1999b,c).

3   A single change in both the mean and the variance of a normal sequence

Consider a set of random variables observed at consecutive equally spaced time points. Suppose that, due to some exogenous factors, the first and second parts of the sequence of random variables operate at two different mean and variance levels. This situation can be represented by the following model :

$$
\begin{array}{ll}
X_i \sim \mathcal{N}\left(x_i \mid \mu_1, \sigma_1^2\right), & i = 1, ..., \tau \\
X_i \sim \mathcal{N}\left(x_i \mid \mu_2, \sigma_2^2\right), & i = \tau + 1, ..., n
\end{array}
\tag{4}
$$

where $\mathcal{N}\left(x_i \mid \mu, \sigma^2\right)$ stands for the usual normal probability density function (p.d.f.) with parameters $\mu \in \Re$ and $\sigma \in \Re^+$. In our opinion, this model is important for hydrometeorological series, since a shift in the mean level often seems to come with a change in variability. Writing $\mu = (\mu_1, \mu_2)$ and $\sigma = (\sigma_1^2, \sigma_2^2)$, the likelihood function resulting from n observations $\mathbf{x} = (x_1, x_2, ..., x_n)$ generated by this model is given by

$$
p\left(\mathbf{x} \mid \mu, \sigma, \tau\right) = \left(\frac{1}{2\pi}\right)^{\frac{n}{2}} \left(\frac{1}{\sigma_1^2}\right)^{\frac{\tau}{2}} \left(\frac{1}{\sigma_2^2}\right)^{\frac{n-\tau}{2}} \exp\left\{-\frac{\tau}{2\sigma_1^2}\left[s_\tau^2 + (\overline{x}_\tau - \mu_1)^2\right]\right\}
\tag{5}
$$

$$
\times \exp\left\{-\frac{n-\tau}{2\sigma_2^2}\left[s_{n-\tau}^2 + (\overline{x}_{n-\tau} - \mu_2)^2\right]\right\}
$$

where

$$
\overline{x}_\tau = \frac{1}{\tau}\sum_{i=1}^{\tau} x_i, \qquad \overline{x}_{n-\tau} = \frac{1}{n-\tau}\sum_{i=\tau+1}^{n} x_i,
$$

$$
s_\tau^2 = \frac{1}{\tau}\sum_{i=1}^{\tau}(x_i - \overline{x}_\tau)^2, \qquad s_{n-\tau}^2 = \frac{1}{n-\tau}\sum_{i=\tau+1}^{n}(x_i - \overline{x}_{n-\tau})^2.
$$

For fixed τ, that expression has the same structure as a product of two normal distributions with two inverted gamma distributions. Assuming conjugacy, suggests a normal-inverted gamma type of distribution to represent prior knowledge about μ and σ :

$$
\begin{aligned}
p(\mu, \sigma) &= \mathcal{N}\left(\mu_1 \mid \phi_1, \lambda_1 \sigma_1^2\right) \mathcal{N}\left(\mu_2 \mid \phi_2, \lambda_2 \sigma_2^2\right) \mathcal{IG}\left(\sigma_1^2 \mid \alpha_1, \beta_1\right) \\
&\quad \times \mathcal{IG}\left(\sigma_2^2 \mid \alpha_2, \beta_2\right) \\
&= \mathcal{NNIGIG}\left(\mu, \sigma \mid \phi, \lambda, \alpha, \beta\right)
\end{aligned}
\tag{6}
$$

where $\phi = (\phi_1, \phi_2)$, $\lambda = (\lambda_1, \lambda_2)$, $\alpha = (\alpha_1, \alpha_2)$, $\beta = (\beta_1, \beta_2)$ are the hyperparameters, and $\mathcal{IG}\left(\sigma^2 \mid \alpha, \beta\right)$ stands for the inverted gamma p.d.f. with parameters $\alpha \in \Re^+$ and $\beta \in \Re^+$:

$$\mathcal{IG}\left(x \mid \alpha, \beta\right) = \frac{\beta^\alpha}{\Gamma\left(\alpha\right)}\left(\frac{1}{x}\right)^{\alpha+1}\exp\left\{-\frac{\beta}{x}\right\}, \qquad x \in \Re^+. \tag{7}$$

Assuming finally prior independence between τ and the other parameters, and that $p\left(\tau\right)$ is the prior discrete distribution for the change-point on the set $\{1, 2, ..., n-1\}$, leads to the joint prior distribution for (μ, σ, τ):

$$p\left(\mu, \sigma, \tau\right) = \mathcal{NNIGIG}\left(\mu, \sigma \mid \phi, \lambda, \alpha, \beta\right) \times p\left(\tau\right) \tag{8}$$

The conjugate distribution assumption, as realistic as others, facilitates the derivation of the posterior distribution. Selecting particular values in (8) for the hyperparameters ϕ, λ, α and β leads us to a different joint prior distribution and allows to take into account a variety of prior beliefs about the studied phenomena. These may come from historical or regional information, even from subjective knowledge. When prior information about the phenomenon is limited, it may be desirable to let the prior knowledge for the unknown parameters be vague (Box and Tiao, 1973). The prior distribution can be turned to a particular form of noninformative density by letting $p\left(\mu, \sigma, \tau\right) \propto p\left(\tau\right) \times \sigma_1^{-2}\sigma_2^{-2}$, i.e. $\lambda_1 = \lambda_2 \to \infty$, $\alpha_1 = \alpha_1 = \beta_1 = \beta_2 \to 0$ in (6).

We shall first evaluate the joint and marginal posterior distributions of μ_1, μ_2, σ_1^2 and σ_2^2 assuming the change-point τ is known. Then, the unconditional posterior densities will be deduced by averaging these distributions over τ. Because of conjugate properties, the conditional joint posterior distribution $p\left(\mu, \sigma \mid \tau, \mathbf{x}\right)$ given τ and the observed data $\mathbf{x}$ also belongs to the class of normal-inverted gamma distributions, but with updated parameters $(\phi', \lambda', \alpha', \beta')$. More precisely, assuming prior independence between the change-point τ and the other parameters, we have as the numerator of the Bayes theorem

$$p\left(\mu, \sigma, \mathbf{x} \mid \tau\right) = p\left(\mathbf{x} \mid \mu, \sigma, \tau\right) \times \mathcal{NNIGIG}\left(\mu, \sigma \mid \phi, \lambda, \alpha, \beta\right) \tag{9}$$

After some algebraic manipulations, the factors involving μ and σ in this last expression are clearly recognizable, up to a normalizing constant, as belonging to a normal-inverted gamma distribution with updated parameters

$$\lambda_1' = \lambda_1 / (1 + \tau \lambda_1) , \qquad \lambda_2' = \lambda_2 / [1 + (n - \tau) \lambda_2] ,$$

$$\phi_1' = (1 - \lambda_1' \tau) \phi_1 + \lambda_1' \tau \bar{x}_\tau, \quad \phi_2' = [1 - \lambda_2' (n - \tau)] \phi_2 + \lambda_2' (n - \tau) \bar{x}_{n-\tau},$$

$$\alpha_1' = \alpha_1 + \tau / 2 , \qquad \alpha_2' = \alpha_2 + (n - \tau) / 2 ,$$

$$\beta_1' = \tfrac{\tau}{2} \left[s_\tau^2 + (1 - \lambda_1' \tau) (\phi_1 - \bar{x}_\tau)^2 \right] + \beta_1,$$

$$\beta_2' = \tfrac{(n-\tau)}{2} \left[s_{n-\tau}^2 + (1 - \lambda_2' (n - \tau)) (\phi_2 - \bar{x}_{n-\tau})^2 \right] + \beta_2.$$

Therefore,

$$p (\mu, \sigma \mid \tau, \mathbf{x}) = \mathcal{NNIGIG} (\mu, \sigma \mid \phi', \lambda', \alpha', \beta') \tag{10}$$

The normalizing constant, i.e. the p.d.f. of the data $\mathbf{x}$ only conditioned upon the change-point τ which makes the Bayes theorem denominator, is denoted $m(\mathbf{x} \mid \tau)$ and can be determined by dividing (9) by (10), and cancelling factors involving μ and σ^2:

$$m(\mathbf{x} \mid \tau) = \left(\frac{1}{2\pi} \right)^{n/2} \sqrt{\frac{\lambda_1' \lambda_2'}{\lambda_1 \lambda_2}} \frac{\beta_1^{\alpha_1} \beta_2^{\alpha_2}}{(\beta_1')^{\alpha_1'} (\beta_2')^{\alpha_2'}} \frac{\Gamma(\alpha_1') \Gamma(\alpha_2')}{\Gamma(\alpha_1) \Gamma(\alpha_2)}. \tag{11}$$

Given τ and $\mathbf{x}$, it is straightforward to show that μ_1 and μ_2 are conditionally distributed as Student t-distributions $ST(\mu_i \mid a, b, c)$, σ_1^2 and σ_2^2 as inverted gamma distributions $IG(\sigma_i^2 \mid \alpha, \beta)$. Integrating the appropriate parameters out of expression (10), leads directly to these conditional posterior distributions of the parameters before and after the change-point. More precisely, we have

$$p(\mu_i \mid \tau, \mathbf{x}, M_3) = ST \left(\mu_i \mid \phi_i', \alpha_i' (\lambda_i' \beta_i')^{-1}, 2\alpha_i' \right), \quad i = 1, 2,$$

$$p(\sigma_i^2 \mid \tau, \mathbf{x}, M_3) = IG (\sigma_i^2 \mid \alpha_i', \beta_i'), \quad i = 1, 2.$$

To draw conclusions regarding the intensity of shifts, it is natural to define the parameters $\delta = \mu_2 - \mu_1$ and $\eta = \sigma_2^2 / \sigma_1^2$. Their conditional posterior distributions can be deduced by simple univariate transformations of variable from the conditional distributions of the original parameters. Given τ and $\mathbf{x}$, η as a Beta distribution of the second kind (Menzefricke, 1981). It can also be shown that $(\alpha'_1 \beta'_1 \eta / \alpha'_2 \beta'_2)$ is distributed as a Fisher distribution with $2\alpha'_2$ and $2\alpha'_1$ degrees of freedom. The posterior conditional distribution of δ is a Behrens-Fisher distribution which cannot be evaluated in closed form. However, it can be efficiently approximated by a Student t-distribution.

Since, according to the Bayes theorem, the joint p.d.f. for $(\mathbf{x}, \tau)$ is just $m(\mathbf{x}|\tau) \times p(\tau)$, the marginal posterior density of the change-point $\tau = 1, 2, ..., n-1$ under model (4) is easily seen to be

$$p(\tau \mid \mathbf{x}) = \frac{m(\mathbf{x}|\tau)\, p(\tau)}{\sum\limits_{\tau=1}^{n-1} m(\mathbf{x}|\tau)\, p(\tau)} \tag{12}$$

$$\propto p(\tau)\, \sqrt{\lambda'_1 \lambda'_2}\, \frac{\Gamma(\alpha'_1)\, \Gamma(\alpha'_2)}{(\beta'_1)^{\alpha'_1}\, (\beta'_2)^{\alpha'_2}}$$

This discrete distribution gives, at time point τ, the posterior probability of simultaneous shift occurrence in both the mean level and variance. To draw conclusions regarding the parameters μ_1, μ_2, σ_1^2, σ_2^2, and the intensity of shifts $\delta = \mu_2 - \mu_1$ and $\eta = \sigma_2^2 / \sigma_1^2$, their marginal posterior distributions must be derived. The corresponding marginal distributions are finite mixtures of the associated conditional distributions weighted by the $n-1$ values of $p(\tau \mid \mathbf{x})$ given by (12). For instance, we have

$$p(\mu_i \mid \mathbf{x}) = \sum_{\tau=1}^{n-1} p(\mu_i \mid \tau, \mathbf{x}) \times p(\tau \mid \mathbf{x}) \tag{13}$$

$$= \sum_{\tau=1}^{n-1} \mathcal{ST}\left(\mu_i \left| \phi'_i, \alpha'_i (\lambda'_i \beta'_i)^{-1}, 2\alpha'_i \right.\right) \times p(\tau \mid \mathbf{x}), \quad i = 1, 2$$

4 Application

Hydro-Québec is a public company that produces, transmits and distributes electricity throughout the province of Québec. It currently operates 54 power plants supplied by 26 large reservoirs. The sites are assembled to form 8 major hydropower systems: St-Laurent, Outaouais, La Grande, St-Maurice, Bersimis, Manicouagan, Outardes and Churchill Falls. For each of these systems, annual energy inflows are evaluated by multiplying the net basin supply of each reservoir into the system by a factor based on the production capacity of the corresponding power plant. Such annual energy inflows are therefore subject to hydrometeorological changes, if any. These data are of high importance for energy planning because some of their statistical characteristics (namely mean and variance) are used as inputs to construct scenarios or to forecast future energy availability.

Consider the time plot in Figure 1 which shows the annual energy inflow for the hydropower system Churchill Falls calculated from 1943 to 1996. These values are expressed in terawatt-hour (TWh).

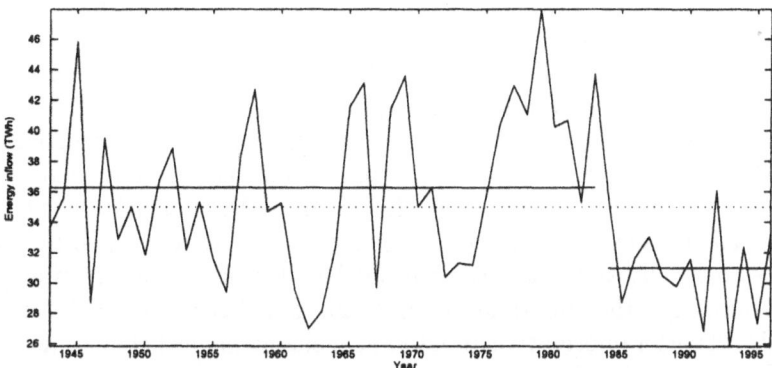

Figure 1: Annual energy inflows at Churchill Falls (1943-1996).

This 92 432 km² watershed is situated in the Labrador, province of Newfoundland (northeast of Québec). Churchill Falls is one of the 8 systems which contribute the most to the total annual energy inflow (approximately 20%). Examining this time series, one may first suspect that an abrupt change in the mean level has occurred around 1983. To meet the energy demand and eventually export generated hydroelectricity, the hydropower company would like to evaluate its safety margin for future energy availability under this new context. Since for this case the mean level has decreased considerably, decisions for the future relying on the overall mean can be dramatically different from anticipations based on the tenth last years. Second, the overall variability of these observations seems to have decreased at the same time. The energy planning system of forecasts depends strongly upon energy inflow variability. For instance, a sharp knowledge of this characteristic is the key point to determine which range of scenarios is needed to make a good decision for reservoir releases and interannual storage strategy.

After examining this series, an engineer will typically base his decision on one of the two possible estimates: evaluating the mean and the variance over the entire period or considering only the last observations, i.e. using more but maybe less relevant information about future realizations or taking into account less but maybe more representative information. A change in the mean and variance may induce important decisions and have large economic consequences; inference about existence and characteristics of such a change can thus be considered as a valuable precaution before developing management rules in water resources systems.

In what follows, Churchill Falls annual energy inflows are analyzed using model (4). First, the prior distributions are specified. Then, inference about the parameters of interest is done using the posterior distributions derived in Section 3.

4.1 SPECIFYING PRIOR DISTRIBUTIONS

The first step in Bayesian analysis is to set up a *full probability model.* That is, in addition to a model for the observable quantities (model (4) for Churchill Falls annual energy inflows), we must represent the prior degree of belief about the unknowns, i.e. the parameters of the models (μ, σ, τ). In Bayesian analysis, specifying a prior distribution for the parameters is part and parcel of the model, with all hypotheses that modelling involves. Prior elicitation is therefore a crucial component of the Bayesian approach. To be fully operational, it is desirable to specify the prior distributions by eliciting knowledge of an expert and/or by using other information than the data itself. An engineer should have valuable prior information about annual energy inflow behavior (subjective knowledge, regional information, etc.).

Experts in Hydro-Québec do not agree about the existence of a shift neither in the mean level nor in the variability of Churchill Falls annual energy inflows. Moreover, if we examine the annual energy inflows for the other hydropower systems, no clear change-point can be identified. Therefore, one can not reasonably favor any year of change, and we assume τ is distributed as an uniform discrete distribution. Note that in this case, the prior expected change-point is 1970, i.e. the mean of a discrete uniform probability distribution on the interval [1943, 1995].

For the other parameters, the prior degrees of belief were assumed to be represented by a normal-inverted gamma type of distribution which p.d.f. is given by (6). The complete specification of prior knowledge about these quantities requires the choice of hyperparameters $(\lambda, \phi, \alpha, \beta)$ in that expression. This was done by eliciting regional information from the 7 other hydropower systems. A regression model was used to predict the average and the variance of the energy inflows for Churchill Falls, before and after 1970. These predictions were then used, along with their standard error, to estimate the mean and the variance of the normal and inverted gamma distributions. Finally, simple systems of equations for the first two moments were solved, yielding to estimated hyperparameter values. A simple regression considering only the generating capacity appeared to be the best regional model for predicting the average and the variance of the energy inflows for Churchill Falls. Details about this approach to derive informative prior distributions can be found in Perreault *et al.* (1999c).

To summarize, the prior mean and standard deviation of each parameter are listed in Table 1.

Table 1: Prior means and standard deviations

	τ	μ_1	μ_2	σ_1^2	σ_2^2	δ	η
Expected value	1970	27.80	27.81	12.88	18.19	0.01	1.47
Standard deviation	16	1.67	2.13	2.73	2.45	2.70	0.36

This table will be used in the next section to show how prior state of belief is updated by the data.

4.2 POSTERIOR ANALYSIS

Expressions given in Section 3, along with the prior distributions specified in the above section, are used to analyze the annual energy inflow of Churchill Falls hydropower complex (Figure 1). Figures 2 reproduces the annual energy inflows of Figure 1 along with the marginal posterior distribution of τ. The marginal posterior densities for the other parameters of interest, together with their corresponding prior density (in dotted line), are presented in Figure 3. We first observe how the data modified or updated prior information by comparing the prior and the posterior distributions. Such graphs, are also useful tools to appreciate the parameters uncertainty.

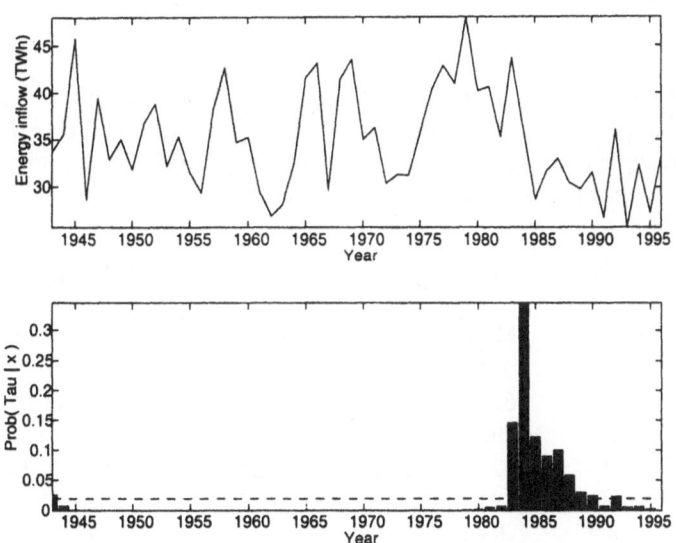

Figure 2: Annual energy inflows and marginal posterior distribution $p(\tau \,|\mathbf{x})$.

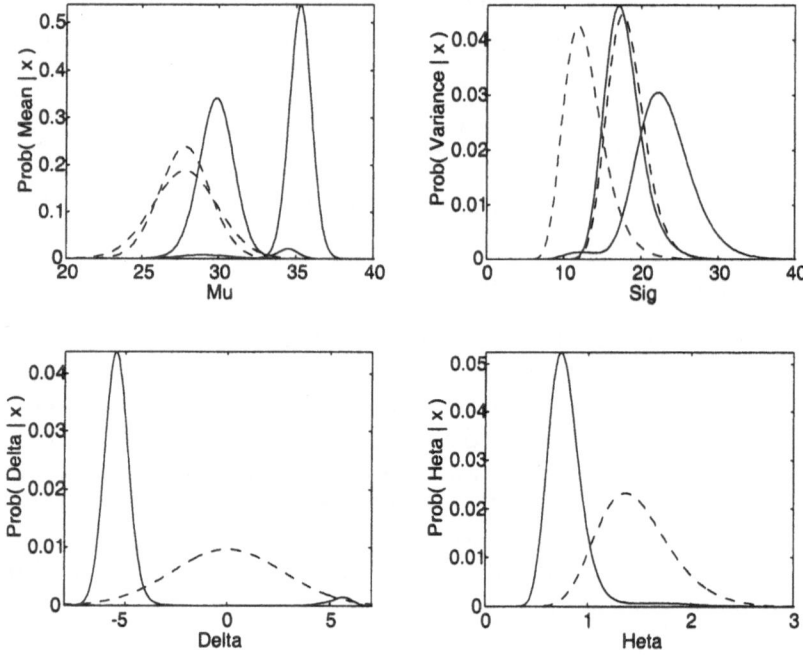

Figure 3: Marginal posterior distributions of μ_i, σ_i^2, δ and η.

Table 2 summarizes estimated posterior quantities (mode, mean and standard deviation). These estimates are to be compared with their corresponding prior values specified in Section 4.1 (Table 1). Table 2 also gives, for each parameter, a symmetric 90% Bayesian credible interval.

Table 2: Posterior moments and credible interval

Parameters	Mode	Mean	Stand. dev.	90% cred. int.
τ	1984	1983.92	7.87	[1982 ; 1989]
μ_1	35.28	35.07	0.74	[33.65 ; 36.37]
μ_2	29.87	30.00	3.48	[27.88 ; 32.13]
σ_1^2	22.33	22.89	3.48	[16.81 ; 29.16]
σ_2^2	17.07	17.81	2.28	[13.92 ; 22.07]
δ	-5.43	-5.07	2.02	[-6.42 ; -4.19]
η	0.73	0.81	0.25	[0.53 ; 1.13]

This analysis indicates first, under the hypothesis of a simultaneous sudden change in mean and variance in the annual energy inflow, that the change-point occurred around 1984 with approximately a 8 years standard deviation. The mean level of energy inflow after the shift seems to have decreased by an amount of 5 TWh with 2 TWh of standard deviation. The credible interval for the intensity of the shift shows that a decrease of 6.5 TWh is still plausible at a 90% credible level. Moreover, this interval do not contain zero which suggests a negative change of at least 4 TWh. However, this observation provides no justification for actions such as "rejecting the no change hypothesis" (the model assumes a change did occur with certainty). Under such hypothesis, the most probable intensity of the shift in the mean level being -5 TWh, the hydropower company would better take action to compensate this potential loss of energy inflow and ensure balance between supply and demand for the next few years. The output of this Bayesian analysis (estimates and uncertainties) can be used to choose among different alternatives (anticipating construction of new power plants, buying electricity, using thermal power station, etc.).

The analysis also indicates that the variance after the change is less that it was before the change (the posterior mode and mean for η are respectively 0.73 and 0.81). Comparing these values to the prior expectation (Table 1), clearly the information stemming from the data contributed considerably to modified the prior state of belief. The credible interval for η shows that an almost 2 times decrease is still plausible at a 90% credible level. But still this interval suggests a much smaller change since it contains one.

5   Discussion

The Bayesian method presented in this paper, can be viewed as an extension of the normal model, and practitioners can perform such change-point analysis routinely using standard statistical toolboxes. This approach can be generalized to other types of p.d.f.'s, for instance the gamma distribution. More precisely, use of probability distributions which belong to the exponential class of p.d.f.'s allows exactly the same line of reasoning based on conjugacy. If prior independence between the epoch of change and model parameters is assumed, the joint posterior distribution is a finite mixture of conjugate distributions. This allows for easy computation of posterior odds, as illustrated by the univariate normal model under the configuration of a single change in both the mean level and the variance. If integration to derive posterior marginals does not lead to expressions in closed form, one can use the Gibbs sampler which is particularly suitable for change-point analysis (Perreault et al., 1999c).

The problem of model uncertainty has not been addressed here. All results herein are conditional upon a given model. Model (4) assumes with certainty that a change in both the mean and the variance has occurred. This statistical reporting framework is essential but incomplete. It forces the hydrologist to choose among the different models to perform a change-point analysis, and to exclude some other kinds of changes. To take into account the "no change hypothesis" and diverse types of changes, one should assign a prior probability to the different alternatives, and consider the change-point study as a Bayesian model selection problem among the various situations that may occur. This perspective for hydrological change-point analysis is explored in Perreault *et al.* (1999b).

Acknowledgments. The authors wish to express their deep gratitude to Professor Jacques Bernier and Lucien Duckstein for very helpful comments. They are also grateful to René Roy and Raymond Gauthier of Hydro-Québec for providing data. This research was supported by la Direction Générale de l'Enseignement et de la Recherche de France of the French ministry of Agriculture (DGER), le fond pour la Formation de Chercheurs et l'Aide à la Recherche du Québec (FCAR), and la Chaire en Hydrologie Statistique Hydro-Québec/CRSNG (INRS-Eau).

References

[1] Berger, J.O. (1985). Statistical Decision Theory and Bayesian Analysis. Springer.

[2] Bernardo, J.M. and Smith, A.F.M. (1994). Bayesian Theory. Wiley, New York.

[3] Bernier, J. (1994). Statistical detection of changes in geophysical series. In Engineering Risk in Natural Resources Management, edited by L. Duckstein and E. Parent, NATO Advanced Studies Institute Series, Kluwer: 159-176.

[4] Box, G.E.P and Tiao, G.C. (1973). Bayesian Inference in Statistical Analysis. Addison-Wesley, Reading, Massachusetts.

[5] Bruneau, P. and Rassam, J.-C. (1983). Application d'un modèle bayesien de détection de changements de moyennes dans une série. Journal des Sciences Hydrologiques, 28: 341-354.

[6] Carlin, B.P., Gelfand, A.E., Smith, A.F.M. (1992). Hierarchical Bayesian analysis of changepoint problems. Appl. Statist, 41: 389-405.

[7] Gelfand, A.E. and Smith, A.F.M. (1990). Sampling-based approaches to calculating marginal densities. J. Am. Stat. Assoc., 85: 398-409.

[8] Hinkley, D.V. (1970). Inference about the change-point in a sequence of random variables. Biometrika, 57: 1-16.

[9] Lecoutre, B. (1997). C'est bon à savoir ! Et si vous étiez un bayésien qui s'ignore. Modulad, 18: 81-87.

[10] Lubès-Niel, H., Masson, J.M., Paturel, J.E. and Servat, E. (1998). Variabilité climatique et statistiques. Etude par simulation de la puissance et de la robustesse de quelques tests utilisés pour vérifier l'homogénéité de chroniques. Rev. Sci. Eau, 3: 383-408.

[11] Menzefricke, U. (1981). A Bayesian analysis of a change in the precision of a sequence of independent normal random variables at an unknown time point. Appl. Statist., 30: 141-146.

[12] Munier B. and Parent, E. (1998). Le développement récent des sciences

de la décision: un regard critique sur la statistique décisionnelle bayési-
enne. In Statistical and Bayesian Methods in Hydrological Sciences, edited
by E. Parent, P. Hubert, B. Bobée, J. Miquel, Thecnical Documents in
Hydrology, 20, UNESCO: 361-398.

[13] Perreault, L., Haché, M., Slivitsky, M. and Bobée, B. (1999a). Detection
of changes in precipitation and runoff over eastern Canada and U.S. using
a Bayesian approach. To appear in Stochastic Environmental Research an
Risk Assessment, 13.

[14] Perreault, L., Bernier, J., Bobée, B. and Parent, E. (1999b). Bayesian
change-point analysis in hydrometeorological time series 2 : comparison of
change-point models and forecasting. Submitted to J. of Hydrol..

[15] Perreault, L., Bernier, J., Bobée, B. and Parent, E. (1999c). Bayesian
change-point analysis in hydrometeorological time series 1 : the normal
model revisited. Submitted to J. of Hydrol..

[16] Savage, L.J. (1972). The Foundations of Statistics. Dover Publications,
New York.

[17] Stephens, D.A (1994). Bayesian retrospective multiple-changepoint identification. Appl. Statist., 43: 159-178.

[18] Tanner, M.H. (1992). Tools for statistical inference: observed data and data
augmentation methods. Lecture Notes in Statistics, 67, Springer-Verlag.

[19] Walley, P. (1991). Statistical Reasonning With Imprecise Probabilities.
Chapman and Hall.

ON THE DISTRIBUTION OF THE MAXIMUM OF A RANDOM SAMPLE

V.F.Pisarenko

International Institute of Theory of Earthquake Prediction and Mathematical
Geophysics, Russian Academy of Sciences, Moscow

1. Introduction

Two problems of extreme value theory are considered. Assessing the first one the
importance is stressed of the direct evaluation of the exact distribution of the
maximum m_n of n independent identically distributed random values with a
distribution function $F(x)$. This evaluation is made in terms of the distribution
denoted $F^n(x)$. Advantages of such an approach are compared with the classical
method providing an asymptotic distribution of normalized maximum m_n. Solving
the second problem a threshold t_n is derived that determines the upper half of the
m_n- distribution. An approximation of the distribution function $F(x)$ can be
facilitated if its range is restricted to $x > t_n$.

2. The exact distribution of maximum values

Let m_n be the maximum of independent identically distributed random values
$x_1,...,x_n$ with a common distribution function $F(x)$. A classical problem of extreme
values consists of the following [1]-[3]: determine normalizing constants a_n, b_n in
such a way that the distribution of $(m_n - a_n)/b_n$ converges to some limit law. It is
necessary as well to describe all possible types of limit laws and to point out classes
of initial distribution functions $F(x)$ belonging to a particular type. This problem
was solved by B.V. Gnedenko in his classical paper [4]. Details on this topic as well
as many applied aspects can be found in [1]-[2].

With the arrival of the computer era this classical problem of extreme
values has lost its applied value. The *exact* distribution function of m_n can be
estimated with the help of a computer for any reasonable value of n. In fact, the
distribution function of $x > t_n$ is equal to

$$P\{m_n < z\} = P\{x_1 < z;...x_n < z\} = F^n(z) = \exp\{n \ \ln F(z)\} \ .$$

(1)

17

M.V. Bolgov et al. (eds.), Hydrological Models for Environmental Management, 17–22.
© 2002 *Kluwer Academic Publishers. Printed in the Netherlands.*

Quantiles z_q of the probability q of exceedance can be found as the roots of equations

$$P\{m_n < z_q\} = q \text{ or } P\left\{F(m_n) < q^{\frac{1}{n}}\right\} = q, \quad 0 < q < 1.$$

(2)

Using eq. (1) gives

$$z_q = z_q(n) = F^{-1}\left(q^{\frac{1}{n}}\right)$$

(3)

where F^{-1} is the inverse function of the function F.

Eqs. (1)-(3) cover almost all applied needs. This fact is not yet fully recognized by the users of the theory of extremes. The principle above can be demonstrated on some examples using some well-known distributions set out below.

1. Frechet distribution

$$F(x) = \exp\left\{-\left(\frac{a}{x}\right)^{\alpha}\right\}, \quad \alpha < 0, \quad x > 0$$

$$z_q = \alpha\left(\frac{n}{\ln\frac{1}{q}}\right)^{\frac{1}{\alpha}}$$

2. Pareto distribution

$$F(x) = 1 - \frac{1}{x^{\beta}}, \quad \beta > 0, \quad x \geq 1$$

$$z_q = \frac{1}{\left(1 - q^{\frac{1}{n}}\right)^{\frac{1}{\beta}}}$$

3. Weibull distribution

$$F(x)=1-\exp\left\{-\left(\frac{x}{a}\right)^{\alpha}\right\}, \quad \alpha>0, \quad a>0, \quad x\geq 0$$

$$z_q = a\left(\ln\frac{1}{1-q^{\frac{1}{n}}}\right)^{\frac{1}{\alpha}}$$

4. Gamma distribution

$$F(x)=\frac{\gamma(\alpha;x)}{\Gamma(\alpha)}=\frac{1}{\Gamma(\alpha)}\int_0^x u^{\alpha-1}e^{-u}du$$

$$P\{m_n<z\}=\exp\left\{n\ \ln\frac{\gamma(\alpha;z)}{\Gamma(\alpha)}\right\}$$

where $\Gamma(.)$ is the gamma function and $\gamma(.,.)$ is the incomplete gamma function.

5. Gaussian distribution

$$F(x)=\Phi\left(\frac{x-a}{\sigma}\right); \quad \Phi(x)=\frac{1}{\sqrt{2\pi}}\int_{-\infty}^x e^{-\frac{u^2}{2}}du$$

$$P\{m_n<z\}=\exp\left\{n\Phi\left(\frac{z-a}{\sigma}\right)\right\} \quad .$$

For the first three cases it can be seen that explicit expressions for quantiles of m_n are obtained. For the fourth and fifth cases the distribution function of m_n is easily calculated for any reasonable value of n. Some computational care (like standard double precision) should be taken in the case of very large n (say, n > 100000). Still the exact quantiles and distribution functions obtained numerically are preferable compared with an asymptotic (i.e. *approximate*) distribution provided by the classical approach.

3. The approximate distribution above a threshold t_n

Another aspect of the problem of extreme values will now be addressed. The quantile z_q with exceedance probability q of the distribution of the maximum m_n is determined from

$$F^n(z_q)=q \quad .$$

$$(4)$$

Suppose a level $q = \overline{q}$ is fixed in eq. (4) and the corresponding quantile $z_{\overline{q}} = z_{\overline{q}}(n)$ is taken as a threshold t_n: $t_n = z_{\overline{q}}(n)$. The threshold t_n corresponds to a quantile of the initial distribution function $F(x)$ with a larger level than $\overline{q}$, namely,

with the level $(\overline{q})^{\frac{1}{n}}$

$$F(t_n) = (\overline{q})^{\frac{1}{n}}$$

(5)

Thus, the $\overline{q}$-tail of the m_n-distribution corresponds to the $(\overline{q})^{\frac{1}{n}}$-tail of $F(x)$. If the domain of interest of the m_n-distribution can be confined by the range $x > t_n$,

then only the $(\overline{q})^{\frac{1}{n}}$-tail of the distribution function $F(x)$ is needed. To be specific, consider the case with $\overline{q} = 0.5$. Then in order to know the "upper half" of the m_n-

distribution only the $(0.5)^{\frac{1}{n}}$-tail of $F(x)$ is needed.

The "lower part" of $F(x)$ corresponding to the range $x < t_n$ can be chosen absolutely arbitrarily. Thus, if one wishes to obtain a parametric approximation for $F(x)$ with the aim of deriving an analytical expression for the "upper half" of the m_n-distribution then it is sufficient to approximate $F(x)$ only for the "tail" $x > t_n$. It is easier to perform this approximation than doing this for the whole range. The choice $\overline{q} = 0.5$ is, of course, somewhat arbitrary.

If, for solving a particular applied problem, interest is only in the behaviour of m_n in the range of very large quantiles (say, $\overline{q} = 0.75$ or even $\overline{q} = 0.90$) then the corresponding threshold t_n is larger and the task of finding an approximation is easier. Thresholds t_n depend on the particular tail function $1 - F(x)$ as shown by eq. (5). Values of the threshold for some popular distributions are given in the Table 1. The last row of this table shows the probabilities for the tails of the initial distributions determining the "upper half" of the m_n-distribution. It is seen that only small parts of the $F(x)$-distribution are used in the m_n-distribution, namely: 5.6% for n = 12; 2.3% for n = 30; 0.7% for n = 100.

Table 1. Values of threshold t_n for different distributions

Distribution	Sample size, n		
	12	30	100
standard exponential	2.88	3.78	4.98
Pareto, $\beta=2$	4.22	6.62	12.03
standard Gaussian	1.59	2.00	2.46
gamma, $\alpha=2$	4.61	5.67	7.05
gamma, $\alpha=3$	6.14	7.34	8.87
Weibull, $\alpha=0.5$, a=1	8.30	14.28	24.75
probability of tail	5.6%	2.3%	0.7%

This conclusion calls for reconsidering the whole problem of parametric approximation of distributions in hydrology: being interested only in the largest values of m_n, there is no need to approximate $F(x)$ for the whole range. Only the tail $x > t_n$ is essential.

By similar reasoning it is possible to derive formulae for thresholds for the case of samples of a random size n. Suppose random values $x_1,...,x_n$ are observed in the time interval T and n is the Poisson random value with magnitude λ. The number of earthquakes in a particular region, for example, can be represented as a random value with the Poisson distribution. The distribution function of m_n can be easily found [5] as

$$P\{m_n < z\} = [\exp\{\lambda T F(z) - 1\}]/[\exp\{\lambda T - 1\}] \quad .$$

(6)

The quantiles of this distribution are derived directly. A simplification can be done in the case $\lambda T >> 1$ when

$$P\{m_n < z\} \cong \exp\{\lambda T[F(z) - 1]\} \quad .$$

(7)

When $\lambda T \to \infty$ the range of interest in eq. (7) is confined by large values of z, so that

$$F(z) - 1 \cong \ln F(z) \quad .$$

(8)

Substituting λT by n (λT is the mean value of the Poissonian random value n) and utilising this latter relation, equations (1) and (7) become similar. Thus, for $\lambda T >> 1$ all the conclusions drawn earlier are valid in the case of random n.

4. Acknowledgment

This work was supported by the grant INTAS 96-1957

5. References

1. Gumbel, E.J. (1958) *Statistics of Extremes*, Columbia Univ. Press, New York.
2. Galambos, J. (1978) *The Asymptotic Theory of Extreme Order Statistics*, John Wiley, New York.
3. Galambos, J. (1994) The development of the mathematical theory of extremes in the past half century, *Probability Theory and its Applications* **39(2)**, 272-293 (in Russian).
4. Gnedenko, B.V. (1943) Sur la distribution limite du terme maximum d'une série aléatoire, *Ann.Math.* **44**, 423-453.
5. Pisarenko, V.F. and Lysenko, V.B. (1997) Probability distribution of maximum earthquake that can occur in a given future time period, *Physics of the Earth* **6**, 15-23 (in Russian).

STOCHASTIC MODEL OF THE RUNOFF FLUCTUATIONS FOR RIVERS WITH A FLOOD FLOW REGIME

A.V. KHRISTOFOROV, G.V. KRUGLOVA and T.V. SAMBORSKI
Dept. of Land Hydrology, Geography Faculty, Moscow State University
Vorobiovy gory, Moscow 119899. tel.:9391001, Fax: +7 095 9328836,
E-mail: ab3304@mail.sitek.ru

1. Introduction

A flood flow regime is typical of rivers over most parts of the world. The particular feature of this regime consists of random annual fluctuations in the number of flood peaks, the dates of their passage, and the height and form of separate peaks. One of the ways of improving methods for hydrological computations is to increase the efficiency of use of the given hydrologic data in taking into account the particular feature of a flood flow regime. For this purpose a stochastic model of fluctuations in runoff during a flood period is presented. It includes a stochastic model of the hydrograph – for a flood season and a stochastic model of the annual fluctuations of its characteristics.

2. Model of the hydrograph for a flood season

The model of the hydrograph describes the temporal variance of water discharge $Q(t)$ during a flood season as a realization of a stochastic process. Time t is considered with a step of one day within an interval $[0, T]$ where $t=0$ is a date of the earliest beginning and $t=T$ is a date of the latest end of a flood season, which are possible for the given river. The model is based on the approximation of the hydrograph $Q(t)$ by the function

$$Q(t) = q_o \, \varphi_o \, (t) + \sum_{j=1}^{k} \, q_j \, \varphi_j \, (t - t_j), \qquad (1)$$

where q_o - is the average discharge of the baseflow for the flood season in the given year;

k - is the number of flood peaks in this year;

$t_1, ..., t_k$ - the dates of passage of the flood peaks;

$q_1, ..., q_k$ - maximum discharges of these floods, independently formed without the influence of baseflow and previous floods;

$\varphi_o \, (t)$ - a dimensionless function of the hydrograph of baseflow;

$\varphi_j \, (t-t_j)$ - a dimensionless function of the hydrograph of flood j, which describes the curve of growth of the jth flood, when $t \leq t_j$, and a recession curve when $t > t_j$.

M. V. Bolgov et al. (eds.), Hydrological Models for Environmental Management. 23–37.

During the development of the model, 60 rivers of the world were examined. These were the rivers of Far East and Nothern Caucasus of Russia, Ukrainian Carpathians, Georgia, Nothern Korea, USA, Algeria, and Surinam. The average duration of hydrological observations was 30 years, so more than 1800 hydrographs were investigated. Basically these were the mountain and semi-mountain rivers with catchment areas between 200 and 25000 km^2 with 5 to 25 high floods during the 3 to 12 months of every year and with a low baseflow. For the most of the investigated rivers, equation (1) can be used with $\varphi_o(t) \equiv 1$ and

$$
\varphi_j(t-t_j) = \begin{cases} 0 & , \quad \text{if } t \leq t_j - \tau_j ; \\ 1 + \dfrac{1}{\tau_j}(t-t_j), & \text{if } t_j - \tau_j < t \leq t_j ; \\ exp\,[-\alpha_j(t-t_j)], & \text{if } t > t_j ; \end{cases} \tag{2}
$$

where τ_j is the duration of the main growth of the jth flood (the discharge of the jth flood grows linearly from date $t_j - \tau_j$ to the date of it's maximum t_j); and α_j is the intensity of it's exponential recession after date t_j .

The hydrograph of the flood period for the River Chenchon near Anchju (Nouthern Korea) in 1957 is shown in Fig.1. The flood period extends from the beginning of May to the end of September, so the numeration of the days begins on 1 May and T is equal to 153 days.

For this year, $k=9$ flood peaks were observed and $q_o = 40$ m^{3}s^{-1}. For this hydrograph the elements t_j , q_j , τ_j and α_j of equations (1) and (2) are given in Table 1.

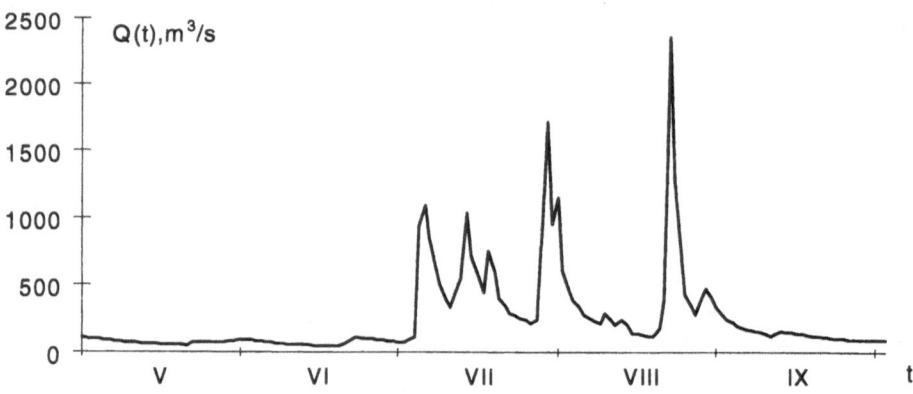

Figure 1. The hydrograph for the River Chenchon near Anchju in 1957.

TABLE 1. The hydrograph elements of the River Chenchon near Anchju in 1957

j	1	2	3	4	5	6	7	8	9
t_j	66	74	78	89	91	100	103	112	119
q_j	1040	860	460	1580	460	140	100	2300	280
τ_j	3	3	2	2	2	1	2	2	3
α_j	0.13	0.20	0.16	0.13	0.20	0.11	0.17	0.25	0.09

The error of approximation does not exceed 50 m^{3}s^{-1} for the floods and 5 m^{3}s^{-1} for the baseflow.

For some rivers the approximation (1) needs more complex functions $\varphi_0(t)$ and $\varphi_j(t - t_j)$. The baseflow is very small in comparison with the floods, so it needs no detailed description. However, for some rivers the function $\varphi_0(t)$ linearly growing in the interval $[t_1, t_k]$ was used. In this it case depends on the flood component of the runoff volume for the whole period. Number k is a characteristic of this volume and correlation between q_0 and k reflects this dependence. For the River Tok (Far East of Russia) $corr(q_0, k)=0.46$, and for the River Lomnitsa (Ukrainian Carpathians) $corr(q_0, k)=0.47$. For functions $\varphi_j(t - t_j)$ formulas of Goodrich or Sokolovsky, or Reits-Crepps can be used. In some cases the recession curve was described by a pair of exponents: one for the first 2 or 3 days and the other for the next [4]. For the investigated rivers the average error of the approximation is no more than 5%. The main variant used was equation (2) for the form of every flood and function $\varphi_0(t) \equiv 1$ for the form of baseflow.

The statistical analysis gave the following results.

1. All elements of every flood peak q_j, τ_j, α_j do not depend on the same elements q_j, τ_j, α_j of another flood peak.

2. The individual maximums $q_1, ..., q_k$ can be considered independent random variables with unified probability distribution for the whole flood season in this case of homogeneity and 2 or 4 functions of distribution for the different parts of the flood period in the case of heterogeneity.

3. The variable τ_j practically depends on q_j and during every flood season the fluctuations of $\tau_1, ..., \tau_k$ can be considered independent random variables with unified probability distribution close to log-normal or gamma with coefficients of variance $C_v = 0.35$-0.50. The mean $E(\tau)$ is close to the time of basin response.

4. The variable α_j does not depend on τ_j and $\alpha_1, ..., \alpha_k$ can be considered independent random variables with log-normal or gamma distribution with $C_v = 0.25$-0.45. The homogeneity of α depends on the homogeneity of q. In some cases it is necessary to consider the correlation between α_j and q_j. For example, for the River Sochi (Northern Caucasus) $corr(\alpha_j, q_j)=0.33$; and for the River Sebau (Algeria) $corr(\alpha_j, q_j)=0.64$. The mean $E(\alpha)$ is close to $1/E(\tau)$.

5. The series of flood peaks during every flood season can be well described by a Poisson process model with the help of function $k(t)$, which determines the mean number of peaks during interval $(0, t)$, where $k(0)=0$ and $k(T)=k_m$ - the mean number of peaks during a whole flood season. In this case the variables $k(t_1), ..., k(t_k)$ behave like the elements of a ranked sample from a uniform distribution over the interval $[0, k]$. If a flood season is relatively homogeneous, function $k(t)$ is linear (Fig 2). For the River Potomac near Hancock (USA) the non-linear curve of function $k(t)$ in Fig.3 demonstrates the intensification of flood activity in a period from January to May.

The variables k, q_0, q_j, τ_j, α_j $(j=1,...,k)$ are the elements of the stochastic model for the flood season hydrograph. In order to obtain the observations of these elements the separation of the observed hydrographs, as in Table 1, is necessary. In the case of n years of hydrological observations n observations for k and q_0 and nk observations for q_j , τ_j , α_j can be used for the statistical estimation. For rivers with relatively homogeneous flood periods and without any big climatic and man-made regular changes, the stochastic process $Q(t)$ can be described with the help of only five functions of probability distribution of the random variables k, q_0, q, τ, α The example of such distributions f for the River Chenchon near Anchju is given in Fig.4 to Fig.8.

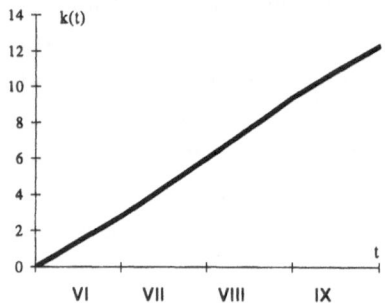

Figure 2. Graph of function $k(t)$ for the River Chenchon near Anchju.

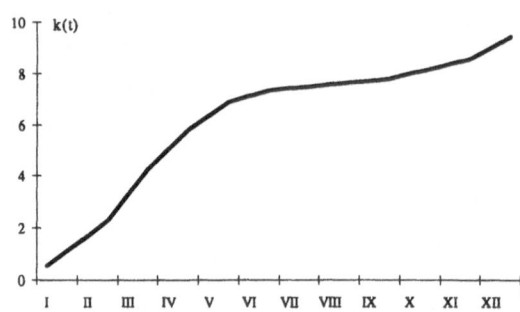

Figure3. Graph of function $k(t)$ for the River Potomac near Hancock.

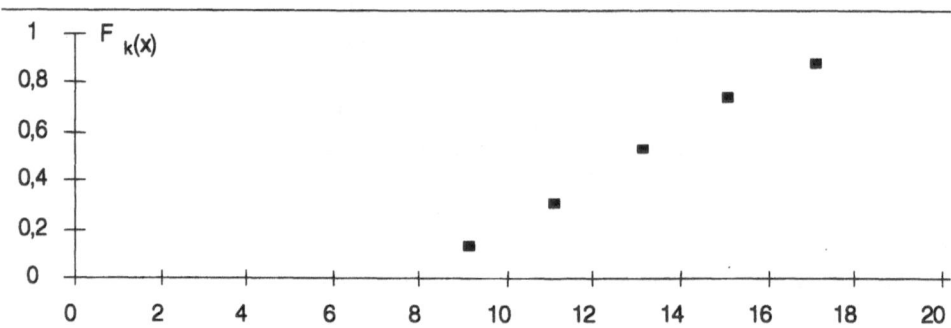

Figure 4. Function of probability distribution F_k (x) of flood peaks number k.

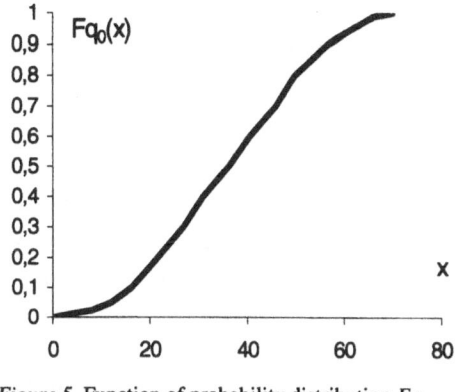

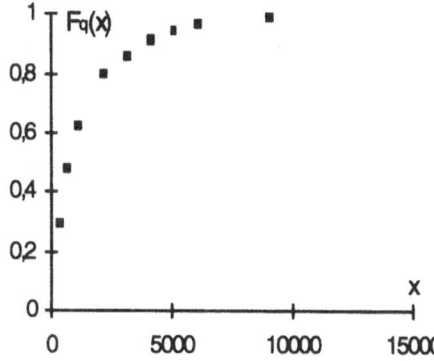

Figure 5. Function of probability distribution Fq_0 (x) for baseflow q_0 .

Figure 6. Function of probability distribution F_q (x) of flood peak individual maximums q.

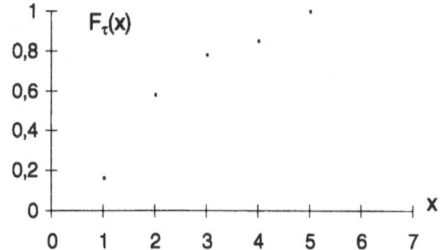

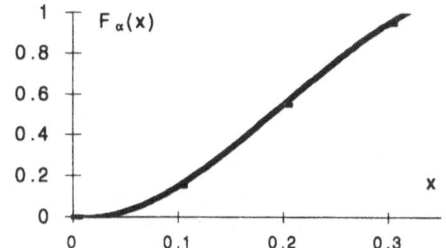

Figure 7. Function of probability distribution F_τ (x) of duration of the main growth τ.

Figure 8. Function of probability distribution F_α (x) of intensity of recession α.

3. Number of flood peaks

The number of flood peaks k is the main characteristic of a flood season. It can be considered as an index of the flood-forming situations of the year. The results of analysis of the dynamics of precipitation and floods confirm this. For the River Lomnitsa near Perevozec (Ukrainian Carpathians, 1490 km^2) the graphs of precipitation $x(t)$ and water discharge $Q(t)$ in 1955 are shown in Fig. 9.

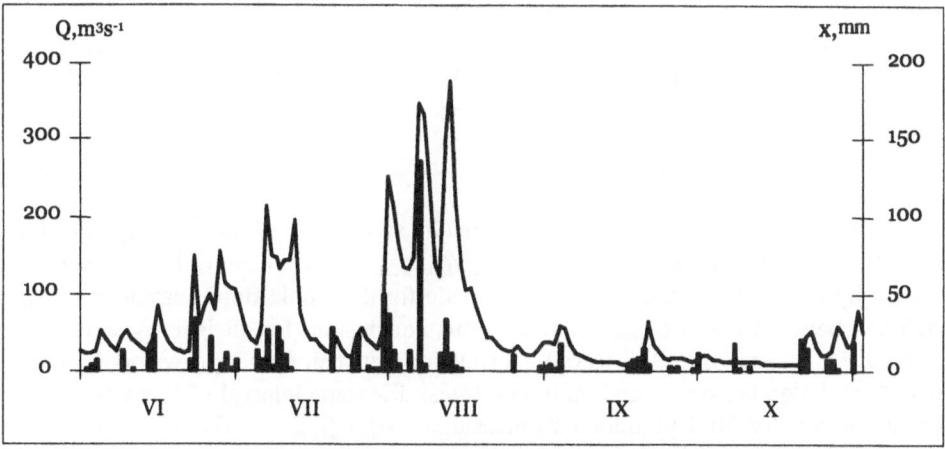

Figure 9. The graphs of $x(t)$ and $Q(t)$ for the River Lomnitsa in 1955.

The time of basin response here is 1-2 days and losses during flood formation depend on the previous moistening so the correlation between maximum flood discharges q and layers x_1, x_2,... of the precipitation of previous days was analysed. This dependence can be described by the formula

$$q = 1.2x_1 + 2.6x_2 + 0.2x_3 - 18 \qquad (3)$$

with coefficient of multiple correlation $R=0.9$. The first three components form the layer of effective precipitation $x_{ef} = 1.2x_1 + 2.6x_2 + 0.2x_3$. The correlation between q and x_{ef} becomes stronger when $x_{ef} \geq 100$mm and it is very important that in every case when $x_{ef} \geq 100$mm the flood with $q \geq 50\text{m}^3\text{s}^{-1}$ appears. So the number of flood peaks k is the number of situations when effective precipitation exceeds some certain critical value, ensuring the formation of a significant flood peak.

For the flood season of every year there is no dependence between the sizes of individual maximums of flood peaks and their number k. In order to motivate this conclusion for some investigated rivers, all n years of observations were split into two groups: n_1 years with $k \leq k_m$ and n_2 years with $k > k_m$, where k_m is a mean number of flood peaks for a given river. For every group the mean number of peaks k_i (i=1,2), the number of observed flood peaks $N_i = k_i\, n_i$, the mean q_i , the coefficient of its variation C_{vi} , and the mean value of water discharge for the whole flood period $Q_{mean}\,(i)$, (i=1,2) were estimated and given in Table 2.

TABLE 2. The characteristics of the flood season for years with high and low flood activity

River	groups	k_i	n_i	N_i	q_i	C_{vi}	Q_{mean}
Tok (Russia)	$k \leq 12$	9.0	25	226	293	1.03	83
	$k > 12$	15.2	19	289	289	1.07	131
Unaha (Russia)	$k \leq 10$	7.1	24	171	117	0.73	34
	$k > 10$	12.2	14	171	99.1	1.07	51.5
Lomnitsa (Ukraine)	$k \leq 6$	3.0	17	52	101	0.93	19
	$k > 6$	8.9	11	98	107	1.15	35
Striy (Ukraine)	$k \leq 12$	7.6	21	160	86.4	1.15	27.3
	$k > 12$	15.9	17	271	108	1.59	56
Svicha (Ukraine)	$k \leq 11$	7.3	17	125	62.5	1.38	16.2
	$k > 11$	13.7	18	247	60.3	1.30	30.3
Uj (Ukraine)	$k \leq 8$	4.5	26	117	46.2	0.84	7.9
	$k > 8$	10.6	16	169	53.8	1.06	20

Years with small and big k have practically similar parameters: $q_1 \approx q_2$ and $C_{v1} \approx C_{v2}$. The proportion k_1/k_2 is close to Q_{mean1}/Q_{mean2} so the number of flood peaks k is a practically equivalent to the Q_{mean} characteristic for the whole flood season. A strong correlation between k and Q_{mean} motivates this conclusion: for all investigated rivers $corr(k, Q_{mean})$ varies from 0.7 to 0.9. Correlation between k and q_0 was considered above. Correlation between k and minimum (mean for some interval of time) discharge is rather strong: for 30-days minimum discharge $corr(k, Q_{min})=0.6-0.8$. The cause is because the intervals between adjacent peaks are reducing with the growing of their number. With the growing of k the probability of very high flood peak grows, but correlation between k and annual maximum Q_{max} is not so strong: $corr(k,Q_{max})=0.4-0.7$. The strong correlation between k and the layer of precipitation during the whole flood season X indicates the defining role of k: $corr(k,X) \geq 0.9$ and in some cases it is more than the correlation between X and mean discharge for the whole flood season.

The defining role of number of peaks was demonstrated by the results of analysis of trends in annual fluctuations of flood season characteristics. The coefficient of correlation between the annual value of the flood season characteristic and the number of the year was used as a measure of the trend and different characteristics of precipitation and runoff were examined. The correlation coefficient between number of the year and annual value of k was the strongest and for 12 rivers of Far East of Russia, Ukrainian Carpathians and Mid-Atlantic of USA these coefficients were statistically reliable. Number of peaks k turned out to be the best indicator of climatic heterogeneity for the rivers with flood flow regime [4].

It is necessary to distinguish a number of rather big flood peaks k from a number of all local maximums of the hydrograph k_0. Number k_o includes all high and low peaks, weakly varies from year to year, insufficiently characterises runoff for the flood period and for sequences of these all, and high and low peaks during a flood season does not correspond to a Poisson model. Number k includes only peaks with individual maximums q which exceed a certain critical value q_{cr}. The optimum value of q_{cr} is defined according to the following considerations:

1) maximum adequacy of the model;
2) maximum conformity of random variable k to Poisson probability distribution;
3) maximum coefficient of correlation between k and mean discharge Q_{mean};
4) q_{cr} must be in the interval from $q(90\%)$ to $q(50\%)$ and from 50% to 90% of all peaks of every flood season must be defined as significant.

For the investigated rivers q_{cr} is close to $2Q_{mean}$. In the case when duration T of the flood period varies from year to year, the variance $D(k)$ exceeds the mean value k (for Poisson distribution $D(k)=k$):

$$D(k) = k + k^2 C_v^2 (T) \tag{4}$$

In order not to lose insignificant peaks in the model the estimation of the probability distribution of q must be based on all $k_0 n$ observed maximums. The Poisson model of flood peak sequences must be described by the function $k_0 (t)$ - a mean number of all peaks during an interval $(0,t)$ of every year:

$$k_0 (t) = (k_0 / k) k(t). \tag{5}$$

4. Model of annual fluctuations

The stochastic model of annual fluctuations of flood season characteristics includes the description of probability distributions of the hydrograph model elements $(k, q_0 , q, \tau, \alpha)$, correlation between these elements, autocorrelation of annual fluctuations of these elements and their climatic and man-made regular changes (trends). If such regular changes are absent or insignificant, the annual fluctuations of model elements and other flood season characteristics are homogeneous. Among the model elements only baseflow q_0 has a statistically reliable autocorrelation. The coefficient of correlation between baseflow for adjacent years r_0 (1) is 0.43 for the River Chita (Russia), 0.44 for Dnestr(Ukraine), and 0.57 for Potomac (USA). Such strong autocorrelation is caused by the inertia of groundwater fluctuations. If Q is some characteristic of flood period flow (mean, maximum or minimum) and $Q^f = Q - q_0$ is its flood component, then the autocorrelation function r_Q (τ) of Q can be determined by the formula

$$r_Q(\tau) = \frac{r_0(\tau)D_0 + r_{f,0}(\tau)\sqrt{D_0}D_f}{D}, \tag{6}$$

where $r_0(\tau) = corr\ (q_0(s-\tau),\ q_0(s))$, $r_{f,0}(\tau) = corr(Q^f(s-\tau),\ q_0(s))$, and D_0, D_f and D are the variances of q_0, Q^f and Q respectively. For $\tau = 1$, the correlation $r_{f,0}$ (1) between base flow q_0 and the flood component Q^f for the previous year is caused by replenishment of the groundwater resources of the present year by the losses of flood flow for the previous year (Q^f partly characterizes it). Annual maximum discharge Q_{max} has no statistically reliable autocorrelation. For some Ukrainian rivers with significant autocorrelation of mean Q_{mean} and 30-days minimum Q_{min} discharges some coefficients are given in Table 3. In columns $r_{k,f}$ (0), $r_{f,0}$ (0) and $r_{f,0}$ (1) the coefficients of correlation between k and Q^f, Q^f and q_0 for the same year and between Q^f for the previous year and q_0 for the present are given for two cases: in the numerator for $Q^f_{mean} = Q_{mean} - q_0$ and for $Q^f_{min} = Q_{min} - q_0$ in the denominator.

TABLE 3. Coefficients of correlation between characteristics of flood period

River	$r_{0,k}(0)$	$r_{0,k}(1)$	$r_{k,f}(0)$	$r_{f,0}(0)$	$r_{f,0}(1)$
Latoritsa	0.59	0.42	0.94	0.48	0.40
			0.53	0.55	0.40
Striy	0.45	0.29	0.93	0.34	0.35
			0.47	0.62	0.37
Prut	0.78	0.34	0.80	0.62	0.24
			0.64	0.56	0.23

Even for Q_{min} correlation between its flood components for adjacent years is not statistically reliable. So for Q_{min} in the case when $\tau = 1$ equation (6) transforms into

$$r_{min(1)} = \frac{r_0(1)D_0 + r(1)\sqrt{D_0 D}}{D_{min}}. \tag{7}$$

The comparison of theoretical and statistical estimates $r_{min}{}^m(1)$ and $r_{min}{}^*(1)$ are given in Table 4 and demonstrate their nearness.

TABLE 4. Estimates of r_{min} (1)

River	D_0*	$r_0(1)$	D_f	$r_{f,0}(1)$	D_{min}	$r_{min}^m(1)$	r_{min}*(1)
Latoritsa	4.1	0.42	6.7	0.40	10.8	0.35	0.30
Striy	14.5	0.38	4.4	0.35	18.9	0.44	0.26
Prut	72.8	0.38	93.2	0.24	166	0.29	0.28

All statistically reliable correlations, including the autocorrelation of q_0, must be considered in the model by means of transformation of correlated variables into normally distributed variables, using linear regression, and subsequent retransformation to the starting probability distribution.

With the help of the Monte-Carlo method, the stochastic model gives the possibility to make theoretical estimation of mathematical expectation (mean) and dispersion (variance) of such important characteristics as the mean Q_{mean}, maximum Q_{max} and minimum mean for 30-days interval Q_{min} discharges. For the simplest variant of the model the hydrograph of every flood can be approximated by means of equation (2) and all elements can be considered independent random variables with homogeneous and uncorrelated fluctuations ("white noise") during a flood season and from year to year. The results of theoretical analysis for the simplest variant of the model are given below.

The flood component Q^f_{mean} of Q_{mean} is proportional to the sum of runoff volumes for all floods and depends on their number k, height q and characteristics of their form τ and α. According to equation (2) the relative volume of every flood is

$$v = \tau/2 + \int e^{-\alpha t}\, dt \approx \tau/2 + 1/\alpha \qquad (8)$$

and the mean value of Q_{mean} is

$$Q_{mean} = q_0 + [(kq)/T][\ \tau/2 + 1/\alpha - \psi(\alpha)]. \qquad (9)$$

The graph of function $\psi(\alpha)$ is given in Fig. 10.

The coefficient of variation of the flood component of mean discharge is given by

$$C_v\,(Q^f_{mean}) = [\sqrt{1 + C_v^2(q)}\sqrt{1 + C_v^2(v)}]\,/\sqrt{k}. \qquad (10)$$

The value Q_{min} is the minimum of mean discharge for a certain interval, for example 30-days. The increase of k conditions the increase of the probability that this certain interval includes the part of one or several floods. The flood component of Q_{min} depends on k and the height, volume and duration of flood peaks. The mathematical expectation of Q^f_{min} is given by

$$E(Q^f_{min}) = E(Q_{min}) - E(q_o) = \bar{q}\,\varphi[\bar{k}, m(v)] \qquad (11)$$

Equation (11) is an approximation arrived at with the help of a simulation experiment. For a 30-day interval the graph of function $\varphi[\ k,\ m(v)\ ]$ is given in Fig.11. If (v) tends to 0, Q_{min} will tend to q_0. If (v) and k are big enough, Q_{min} is close to Q_{mean}.

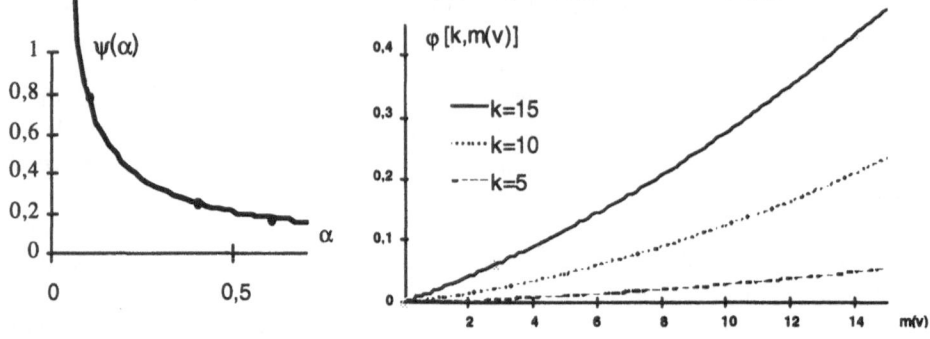

Figure 10.The graph of function $\psi(\alpha)$.　　　　Figure 11. The graph of function $\varphi[\ k,\ m(v)\ ]$

The coefficient of variation $C_v\ (Q^f_{min})$ of the flood component of Q_{min} increases with the reduction of k and increase of q. If the relative volume and duration of the flood peaks are rather big (big τ and small α), $C_v\ (Q^f_{min})$ approximately does not depend on q. For a 30-day interval and for some cases the values of $C_v\ (Q^f_{min})$ are given in Table 5.

TABLE 5. The values of $C_v\ (Q^f_{min})$ for some cases

	τ=2 and α=0.6			τ=3 and α=0.4			τ=6 and α=0.1		
q	50	100	300	50	100	300	50	100	300
k									
5	1.12	1.63	2.21	1.12	1.86	2.56	1.50	1.57	1.55
10	0.78	1.08	1.33	0.95	1.19	1.31	0.87	0.86	0.82
15	0.73	0.81	0.87	0.70	0.77	0.81	0.61	0.62	0.60

The annual maximum discharge Q_{max} includes the baseflow q_0 and depends on the number k and height q of the floods peaks. If the influence of flood superposition is ignored (it is possible when k is small and α is big) then Q_{max} - q_0 can be considered the maximum among k variables q_1 , ..., q_k. Really, Q_{max} depends on the recession of the floods prior to the maximum. The mean Q_{max} is given by

$$E(Q_{max})=q_0 + q\mu(k,\alpha) \qquad\qquad (12)$$

The graph of function $\mu(k,\alpha)$ is given in Fig.12. The coefficient of variation $C_v(Q^f_{max})$ of the flood component of Q_{max} depends on k and α and the graph of its dependence is given in Fig. 13.

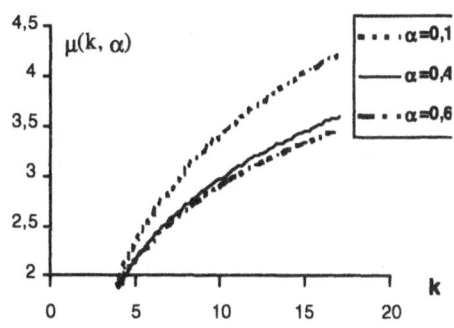

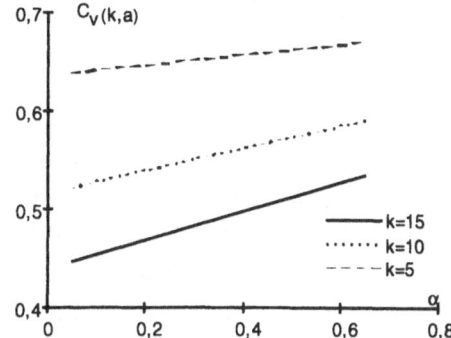

Figure 12. The graph of function $\mu(k, \alpha)$

Figure 13. The graph of the dependence of C_v on k and α

The results of theoretical estimations with the help of formulas and graphs given above are compared with the results of statistical estimations for the Rivers Lomnitsa (Ukrainian Carpathians) and Unaha (Far East of Russia) in Table 6.

TABLE 6. The theoretical and statistical estimates of the characteristics of flood period

River	estimate	Q_{mean}	C_v	Q_{min}	C_v	Q_{max}	C_v
Lomnitsa	theor..	24.2	0.32	7.80	0.40	310	0.56
	Stat..	26.3	0.46	7.64	0.40	276	0.60
Unaha	theor..	38.0	0.35	11.0	0.41	341	0.54
	Stat.	40.4	0.33	13.3	0.64	302	0.45

This comparison demonstrates a good possibility of the theory. In the case of more complex variants of the model it is impossible to obtain any formulas and it is necessary to use the Monte-Carlo method.

Possible climatic change may cause a trend in annual fluctuations of k. The strong correlation between k and volume or layer of precipitation during a flood season can be used for the correction of expected values of k on the basis of the forecast of regular change of precipitation. The direct man-made changes of flood formation (water resource management) can be considered by the water-balance method, and the indirect man-made influences (forest change, melioration, urbanization,...) by regional empirical formulas [4].

5. The reliability and efficiency of the model

A stochastic model of runoff fluctuations during a flood season can be used in hydrological computations by means of the creation of a great number of artificial probable realizations of a process $Q(t)$ during the flood season of the years of certain period with the help of the Monte-Carlo method, allowing computation of certain characteristics of the flood period for every year and estimation of its probability distribution. Reliability of the model depends on its adequacy and statistical stability. The error in the result of the model computation contains two components. The first one is the systematic error of approximation caused by the inevitable difference between reality and the hypothesis of the model. The second one is caused by the random errors of statistical estimations of model parameters based on limited hydrological data. The more complex variants of model have the better possibilities of identical description of real flow fluctuations and have the smaller systematic error of approximation. However, such models include more parameters and statistical estimation of them reduces the statistical stability and increases the second component of error. Such contradiction defines an optimum degree of model complexity for each specific case.

The adequacy of the model is tested currently during it's development for a specific river by statistical testing of the hypotheses of the model. In order to test the adequacy of the model as a whole it is necessary to compare the result of computations executed with the help of the model and with the help of traditional statistical analysis of annual observed river flow characteristics. In Table 7 the results of the model(^{m}) and traditional computations are given for 1% quantiles of annual maximum discharge Q_{max} m^3s^{-1}, volume of river flow for a whole flood season W km^3, annual necessary volume V mln.m^3 of flood-control reservoir with fixed regulated discharge Q_r (the direct problem of river flow regulation) and Q_r m^{3}s^{-1} with fixed V (the opposite problem of regulation).

TABLE 7. Comparison of the results of computations

River - country	Q_{max}	Q^m_{max}	W	W^m	V	V^m	Q_r	Q^m_r
Chita - Russia	450	700	0.68	0.72	1.04	1.12	72	78
Tok - Russia	2460	2220	2.71	2.58	2.03	2.2	200	180
Unaha - Russia	774	835	0.91	0.91	0.75	0.70	180	192
Lomnitsa - Ukraine	846	884	0.81	0.65	0.60	0.49	360	290
Chenchon - Korea	13800	11800	3.25	3.5	0.58	0.58	1050	1060
Sebau - Algire	2620	3160	2.12	2.16	1.91	1.92	164	180

The comparison of the results of computations demonstrates the good possibility to describe the annual river flow fluctuations by the model.

The statistical stability of the model must be tested with the help of the Monte-Carlo method. For the simplest variant of the model and different values of its parameters, 500 artificial realizations of the process $Q(t)$ for n years were generated. The Poisson probability distribution for k and log-normal distribution for q_0 , q, τ, α were used. The parameters of the model (the mean m and coefficient of variation C_v of every model element) were estimated for each realization and 500 estimates were used for determination of the relative error (coefficient of variation of the estimate) for each model parameter. Some results are given in Table 8.

TABLE 8. The relative errors of model parameter estimation (%)

Variant		k	$m(q_0)$	$C_v(q_0)$	$m(q)$	$C_v(q)$	$m(\tau)$	$C_v(\tau)$	$m(\alpha)$	$C_v(\alpha)$
k=9	n=15	8.5	7.7	18.9	8.5	7.9	3.6	6.0	2.5	5.6
	n=30	6.1	5.4	13.5	6.3	5.6	2.4	4.3	1.9	4.0
k=16	n=15	6.5	7.6	18.7	6.5	6.1	2.6	4.5	2.0	4.3
	n=30	4.4	5.5	13.3	4.6	4.2	1.9	3.3	1.5	3.1

Except for the not so important model parameter $C_v (q_0)$, the errors of all the estimates do not exceed 10% in the case of only $n=15$ years of hydrological observations, so the statistical stability of the model is sufficiently high.

The efficiency of the model was tested by comparison with the results of computations fulfilled with the help of the model and with the help of present normative methods. Tests were executed by the Monte-Carlo method. Different quantiles of the mean Q_{mean}, maximum Q_{max} and 30-day minimum Q_{min} discharges and the necessary volume of flood-control reservoir V with $Q_{\hat{a}}=0.3Q_{mean}$ were compared for a case of $n=30$ years of hydrological observations. Some results are given in Table 9.

TABLE 9.The relative errors of hydrological computation by normative methods (e_n) and by the model (e_m)

k	ε	Q_{max} (1%)	Q_{max} (5%)	Q_{min} (90%)	Q_{min} (95%)	Q_{mean} (80%)	Q_{mean} (90%)	V (1%)	V (5%)
9	ε_n	21.0	16.1	9.2	10.6	6.8	8.2	15.0	11.4
9	ε_m	12.3	10.3	6.1	6.6	4.7	4.8	8.9	6.5
16	ε_n	19.6	14.7	13.3	15.7	7.6	9.1	12.4	9.7
16	ε_m	8.6	7.5	6.9	7.5	4.9	5.1	6.4	6.6

The results of the test demonstrate that the model allows the accuracy of the computations to be increased by a factor of 1.5 - 2 and more.

The model is intended for the hydrological computations of any characteristics of river flow during a flood season for weakly studied rivers in cases where a lack of hydrometeorological observations don't allow use of models of river flow formation.

6. Example of application

The application of the model is not limited by the traditional problems of hydrological computations. This is confirmed by an example of model application for the analysis of annual flow fluctuations of rivers of the Mid-Atlantic region of the USA. For this region the systematic increase of river flow from the beginning of the 1970s was

36

discovered. The River Potomac is used here as an example. The other investigated rivers of the Mid-Atlantic region (Susquehanna, Delaware, Shenandoah, Rappahannock and James R.) have approximately the same situation. For the River Potomac at Hancock the graphs of annual fluctuations of the mean annual discharge Q_{mean}, baseflow q_0 and number of rather big flood peaks k with q more than $200m^3s^{-1}$ ($7000ft^3s^{-1}$) are given in Fig. 14. The river basin ($10550km^2$) is situated basically in the Appalachian Mountains. The maximum of flood activity is in March (2 rather big floods in the mean) and in April (1.55 floods), and the minimum flood activity is in July-September (0.15 floods per month).

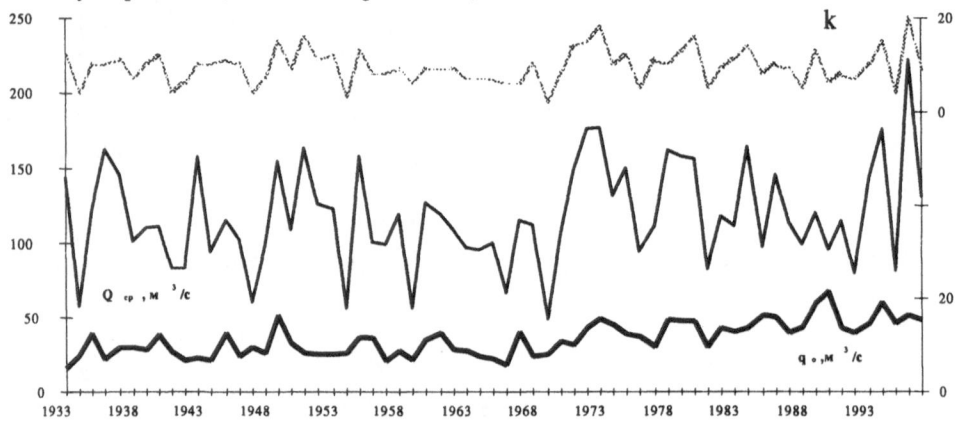

Figure 14. The graphs of annual fluctuations of Q_{mean} , k and q_0 for the River Potomac at Hancock

For the periods from 1933 to 1969 (n_1 =37) and from 1970 to 1997 (n_2 =28) the estimates of the mean X and standard variation S of k, Q_{mean} , q_0 , Q_{max} and q are given in Table 10.

Table 10. The river flow characteristics of compared periods

	$\bar{x}_1$	$\bar{x}_2$	S_1	S_2	τ
k	8.64	10.54	3.16	4.08	2.10
Q_{mean}	108.4	131.0	31.0	35.1	2.69
q_0	9.2	14.4	2.4	2.8	7.90
Q_{max}	1682.8	1598.9	1260.6	950.7	0.50
q	585.8	557.9	601.6	517.2	0.19

The comparison of the estimates S_1 and S_2 with the help of Fisher's criteria demonstrated that the hypothesis of variance constancy can be accepted with risk level 5% for all five characteristics. The estimates $\bar{x}_1$ and $\bar{x}_2$ were compared with Student's criteria with measure τ

$$\tau = \mid \bar{x}_1 - \bar{x}_2 \mid \sqrt{(n_1 n_2)/n} \sqrt{n - 2} \ [(n_1 - 1) S_1^2 + (n_2 - 1) S_2^2]^{-1/2} . \quad (13)$$

The values of τ are given in Table 10 and demonstrate that in a case of 5% risk level the increase of k, Q_{mean} and q_0 is statistically reliable. The increase of base flow q_0 is the strongest and it is possible to establish the trend as the coefficient of correlation between annual q_0 and number of the year is 0.68. However, the reliable changes of annual fluctuations of maximum discharge are absent.

According to [5] the significant changes of the conditions of river flow formation for the Potomac river basin are absent. The results of statistical analysis of q, τ and α variability demonstrate that for these elements of the model the parameters of the probability distribution estimated for the compared periods (1933-1969 and 1970-1997) are practically similar. For q such comparison is given in Table 10. So the climatic change must be the cause of the increase of mean and minimum river flow of Mid-Atlantic rivers.

According to [1] the statistically reliable change of annual precipitation is absent, but the annual volume of precipitation is not the representative characteristic of flood activity in this region. The evident contradiction between the strongly uneven distribution of flood peak frequency within a year (Fig. 3) and very even distribution of precipitation confirms this. It is very important that according to [2] the frequency of winter thaws and hard rains in summer and autumn increased from the beginning of the 1970s. That is why the mean number of significant flood peaks k_m increased by a factor of 1.22 (Table 10). This number has a strong influence on the annual flow: for the River Potomac the coefficient of correlation between k and Q_{mean} is 0.85. According to formula (10), for the Potomac the increase of the mean k from $k_m = 8.6$ to $k_m = 10.5$ must give the increase of the flood component of the mean annual discharge $Q_{mean} - q_o$ as 15.2m³s⁻¹. According to Table 10 the real increase is 17.7 m³/s. The difference between theoretical and statistical estimates is not beyond the bounds of probable values of random errors of statistical estimation.

The theory of the model can explain the absence of significant change in the fluctuations of maximum discharge Q_{max} : according to formula (12) and the graph of Fig. 12 the increase of the mean number of flood peaks k_m from 8.64 to 10.54 with mean $\alpha=0.3$ for the Potomac must give only 9% increase of Q_{max} and, according to [3] in this case, the possibility of Student's criteria is not enough to reveal such a small change.

The baseflow has the strongest change: for the Potomac the mean q_0 increased 1.57 times. According to [5] 77% of the basin is covered with forests and infiltration is rather high (85mm per year). The most part of this water replenishes the groundwater resource which determines the value of the baseflow. The increase of frequency of flood-forming rains in summer and autumn caused the increase in mean number of flood peaks during July-September to 1.4 times - more than the increase of flood peak number over the whole year. The increase in number of floods during this period caused the increase of infiltration, groundwater and, as a result, baseflow. The increase of flood activity especially during summer-autumn explains the trend of baseflow annual fluctuations over the last 30 years.

7. References

1. *IPCC* (1996) Climate Change 1995: The Science of Climate Change. Cambridge Univ. Press, 572pp.
2. Karl, T.R., Knight, R.W. and Plummer, N.P. (1995) Trends in high-frequency climate variability in the twentieth century. *Nature*, 377, 217-220.
3. Khristoforov, A.V. (1993) *Reliability of river flow computations*. Moscow Univ. Press, 168pp.
4. Khristoforov, A.V., Kruglova, G.V. and Samborski T.V. (1998) *Stochastic model of river flow fluctuations during a flood season*. Moscow Univ. Press, 146 pp.
5. Slack, J.R., Lumb, A.M and Landwehr, J.M. (1993) Hydro-Climatic Data Network (HCDN): Streamflow Data Set, 1874-1988. *USGS Water Resources Investigations Report 93-4076* (CD-ROM).

A STUDY OF THE SYNCHRONICITY OF ANNUAL RUNOFF FLUCTUATIONS

V.M. EVSTIGNEEV and T.A. AKIMENKO
Department of Land Hydrology, Geography Faculty, Moscow State University
Vorobiovy gory, Moscow, 119899
e-mail: taa@hydro.geogr.msu.su

1. Introduction

The spatial structure of long-term fluctuations of river runoff has been studied for a long time in connection with research on the genesis of these fluctuations and attempts to establish their global and large-scale laws. One of the directions of this research is the identification of areas with synchronous and asynchronous fluctuations of annual runoff. This topic is also connected to such practical tasks as: a) estimation of the reliability of statistical parameters of time series and their adjustment to the long-term periods, b) evaluation of the spatial correlation in data as a factor for reducing the reliability of maps and empirical dependencies; and c) estimation of the effect of runoff asynchronicity on the common operation of large hydropower and other water use systems.

Various less rigorous approaches were used in early works to identify areas with synchronous runoff fluctuations. Attempts were made to use information contained in the correlation matrix, applying methods such as factor and principal component analysis (PCA) and decomposition into empirical orthogonal functions (EOF). These approaches firstly did not lead, as a rule, to objective conclusions, and secondly, gave ambiguous results which were difficult to explain. That is why there is a need to develop other approaches able to cope with the task of identifying areas with synchronous runoff patterns, and which combine objectivity of results and simplicity of algorithms. A number of studies on this topic were carried out in the 1970s and 1980s at the Department of Hydrology of Moscow University [2,3]. Later, different ways to approach the problem have been based on the use of cluster analysis which simplifies handling of large size correlation matrices arising from simultaneous consideration of numerous time series.

A simple and reliable approach for identifying areas with synchronous runoff fluctuations is described in this paper and applied to a case study which employs data from the European part of the former USSR. The approach has its background in a basic algorithm developed for this purpose by Bykov et al. [3].

M.V. Bolgov et al. (eds.), Hydrological Models for Environmental Management, 39–50.
© 2002 *Kluwer Academic Publishers. Printed in the Netherlands.*

2. An algorithm for identification of synchronous runoff areas

The algorithm for identification of synchronous runoff areas applied here can be briefly described as follows. Let $\left\| r_{i,j} \right\|$ denote a correlation matrix of time series at N observation stations, in arbitrary order. A critical level of pair-wise correlation r_{cr} is defined. It is established either based on, for example, the reliability of estimation of r_{ij}, or by means of trial and error, looking at the sensitivity of results due to a consecutive change in r_{cr}.

Each observation station is attributed to a group in the following way:

a) Initially a certain station is joined with the one station having the highest correlation coefficient exceeding r_{cr}. If there is no such station, the initially considered station forms a degenerated group of one element.

b) In a second step, the observation station having the maximum sum of correlation coefficients (each of them should be above r_{cr}.) with the two stations in the group joins it. If there is no such station, a group of two elements is formed.

c) The fourth, fifth etc. observation stations are joined with the group in a similar way, as long as there is at least one observation station having a correlation coefficients not lower than r_{cr} with each of the previous ones.

N groups are formed in this way, one for each station. The observation stations, having a completely similar group, are joined together and form clusters of a region. So a set of clusters is formed. These clusters have the following properties [1]:

a) the set of stations of each cluster is unique;

b) all mutual correlations for all pairs of stations within a group is not lower than r_{cr};

c) the average of all mutual correlation coefficients within a cluster is higher than with stations of any other cluster.

d) the formed clusters do not break up when r_{cr} is decreased and their set of stations is expanded by association of single observation stations or by unification of clusters

Basically, the described procedure is a standard way of constructing a pattern recognition system with self-training. There are objects with their individual characteristics, but it is not specified to what classes these objects belong. Neither the characteristics of these classes nor their number are known. The purpose of self-training is to find clusters and their number on the basis of the initial data and target conditions of objects' association into classes. In terms of the pattern recognition theory the identification of regions with synchronous runoff may be formulated as follows. There are N objects (items of river basins), described by an ordered set of attributes $(z_1,..., z_n)$ derived as standardized annual runoff time series $Q_1, ..., Q_n$:

$$z = \frac{Q - \overline{Q}}{\sqrt{\sigma_q}}; \quad \overline{z} = 0; \quad \sigma_z = 1. \tag{1}$$

The number of classes and their description in terms of attributes z are not known for the specific situation. The purpose of classification is to combine objects (e.g. observation stations) into clusters for some given objectives. These objectives follow from the tasks for identification of the areas with synchronous runoff: the correlation coefficients between time series of stations which are included in one class should be maximal and not lower than some threshold r_{cr}. Such conditions, in turn, define a measure of proximity of objects in the attribute space $(z_1, ..., z_n)$:

$$r_{ij} = \frac{1}{n} \sum_{m=1}^{n} z_{i,m} z_{j,m}, i, j = 1,...,k .$$ (2)

This correlation coefficient has a linear relationship with the squared Euclidean distance between objects:

$$\Delta_{ij}^2 = \sum_{m=1}^{n} \left(z_{i,m} - z_{j,m}\right)^2; \; d_{ij}^2 = \frac{\Delta_{ij}^2}{n} = 2\left(1 - r_{ij}\right), \; r_{ij} = 1 - 0.5 d_{ij}^2 .$$

(3)

All possible sets of objects in a task of synchronicity can be geometrical interpreted as a set of points with co-ordinates $\{\xi_1, ... ,\xi_n\}$ included in an n-dimensional unit hypersphere, as $-1 \le r_{ij} \le 1$, $i.e.:$

$$0 \le d_{ij}^2 \le 2, \text{ where } d_{ij}^2 = \sum_{m=1}^{n} \left(\xi_{i,m} - \xi_{j,m}\right)^2 \text{ and } \xi = \frac{z}{\sqrt{n}} .$$ (4)

The geometrical interpretation in two-dimensional attribute space (ξ_1, ξ_2) facilitates study of the algorithm's functioning in different situations. Numerous experiments showed that dense non-overlapping clusters could be effectively allocated in attribute space, applying the algorithm with $r_{cr} > 0$. The density of objects in these clusters is adjusted by assigning different values to r_{cr}. When r_{cr} decreases the number of elements in the cluster extends by association of single objects and groups of only two objects and some clusters are joined together. But a decrease in r_{cr} does not always result in integration of clusters. In any case, the distances between centres of clusters are greater than the average distance between objects of one cluster. If the choice of r_{cr} is not rigidly stipulated by the task and plays the role of an algorithm parameter, it is possible to consider the dynamics of object association (attributing different values to r_{cr}) and to choose the most suitable variant, based on a particular purpose for classification or regionalisation over a territory. For a high density of objects in the attribute space, the clusters formed sometimes keep their structure stable after having obtained a certain density. In this case, if the geographical interpretation of the classification is good, it is possible to pass to semantic formulation of the results. If such an interpretation is lacking, the problem is how to further associate clusters. Obviously other rules should be formulated. When studying synchronicity of runoff fluctuations this problem is not very difficult, as the purpose of the classification is precisely formulated: to find the sets of the objects with the highest correlation and not lower then the given level r_{cr}.

The identification of clusters gives essential information about spatial correlation but, strictly speaking, is not a solution to the task of regionalisation. The identification of regions serves the purpose of creation of a geographical system of diagnostics. If an ungauged basin is attributed to a certain region, we assume (with a greater or smaller degree of reliance) that this object has correlation of not less then R_{min} with other objects in this region and the average correlation coefficient between them is higher than with objects of any other region. The regionalisation task includes some important stages. It is useful to discuss them based on a concrete example together with the problems that arise.

3. A case study for the European part of the former USSR.

The area studied is situated in the European part of the former USSR. The data set consists of 91 annual discharge series for 1945-1985 with an average observation period of 38 years. On average it was possible to estimate 36 pairs of correlation coefficients. The choice of time series was made using the following criteria: a basin area from 9000 up to 40000 km², river basins not nested, and at least 30 years of observations. The observation stations are shown in Figure1.

Definition of the threshold correlation r_{cr}

If the level of a correlation is not rigidly stipulated by the task, it is reasonable to solve this problem with a trail and error procedure, with respect to the degree of detail required in regionalisation. In our example, the result of regionalisation for a value r_{cr} ≥0.4 is too detailed and no fewer less than 19 clusters are found. A lowering of r_{cr} to 0.2 results in a more feasible pattern with 12 clusters. A further lowering of the threshold level r_{cr} is not justified because firstly, it does not significantly modify the degree of detail achieved in the regionalisation, and secondly, a doubt arises concerning the accuracy of estimation for individual samples for $r_{ij} < 0.2$ to reliably evaluate the correlation between the joint observations ($\overline{n} = 36$ years). In fact the t-test significance level $\alpha = p(r \geq 0.2 / \rho = 0) = 12\%$, i.e. the risk of an erroneous statement about a direct correlation at evaluation $r_{ij} = 0.2$ is already rather large. However, this factor should not be exaggerated for setting r_{cr}. In our example, the time series with an average correlation in clusters between 0.566 and 0.806 were united at $r_{cr} = 0.2$. The empirical distribution function of all 355 estimated correlation coefficients between objects of clusters r_{ij} is shown in Table 1 together with t- distribution probabilities $p(r / \rho = 0)$.

TABLE 1. Common characteristic of correlation within clusters

r	0.75	0.60	0.50	0.40	0.30	0.25	0.20
$P(r)$, %	25	55	75	90	98	99	100
$\alpha = p(r/\rho=0)$	-	-	0.0005	0.003	0.03	0.07	0.12

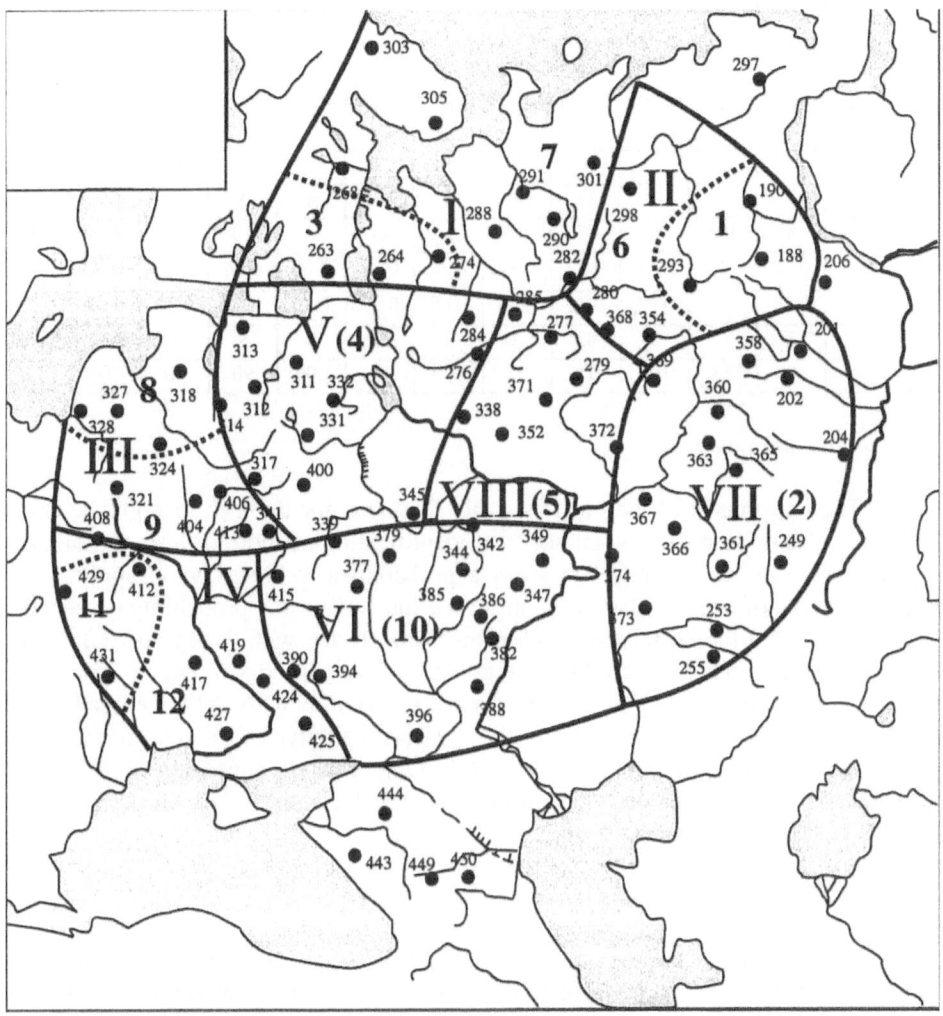

Figure 1. Regionalization of the former USSR territory for synchronous annual runoff fluctuations (1965-1985).

Legend: ● 444 –basin centres, ••••• **10** – region clusters, ▬▬▬ **IV** regions

Examining the joint probability distribution for these data, it is easy to get an estimate of the part of cases with a correlation $\rho_{ij} \leq 0$ in the common set at $r_{cr} = 0.2$, namely: $P(r_{ij} \geq 0.2 / \rho \leq 0) \leq 0.65\%$. In the clusters obtained, the above mentioned risk does not exceed 0.5 % and can be ignored. Thus, it is possible to conclude that the river basins with a direct correlation of runoff time series are joined into a set of clusters, and the major part of intra-cluster correlation coefficients exceeds 0.5. Characteristics of the individual clusters are shown in Table 2. The following abbreviations are used in the table: N - number of objects, N_r- number of correlation coefficients r_{ij}, $\bar{r}$ and r_{min} - mean and minimum value of correlation coefficients between objects, and $N_{(r>0.5)}$ - number of coefficients $r_{ij} > 0.5$, $d_{(r>0.5)} = N_{(r>0.5)} / N_r$.

TABLE 2. Individual characteristics of cluster correlation

cluster	1	2	3	4	5	6	7	8	9	10	11	12
N	3	15	3	11	7	3	7	4	5	15	3	5
N_r	3	105	3	55	21	3	21	6	10	105	3	10
$\bar{r}$	0.70	0.59	0.81	0.62	0.71	0.78	0.57	0.69	0.68	0.60	0.73	0.70
r_{min}	0.59	0.21	0.74	0.33	0.40	0.68	0.37	0.55	0.42	0.22	0.65	0.52
$N_{(r>0.5)}$	3	77	3	43	20	3	12	6	9	72	3	10
$d_{(r>0.5)}$, %	100	73	100	78	95	100	57	100	90	69	100	100

The regionalisation task

The importance of the clusters is not the same for the task of regionalisation. The small clusters do not allow an accurate identification of more or less vast areas of regions. Let's look at this problem from a probabilistic point of view applying the binomial scheme of events. The 95% confidence interval for the probability of rejection of the hypothesis of a correlation within the 3-rd cluster with 3 objects and with 3 correlation coefficients r_{ij}, exceeding the given level R_{min} is: $0 < p\ (A) < 0.708$. This result makes it doubtful for a dominance of correlation coefficients not lower than R_{min} in the considered region. For the region with 4 objects a similar calculation gives a more convincing result, namely $0 < p\ (A) < 0.459$. Only clusters consisting of 5 or more objects allows regions to be detected with confidence: for 5 objects $p\ (A) < 0.31$, for 6 - $p\ (A) < 0.22$, for 10 - $p(A) < 0.14$ etc.

Integration of clusters

In our example the integration task is actually only for the small clusters and should be solved using criteria similar to those used for the identification of clusters. These criteria are rather obvious:
1) find the maximum of the average value of inter-cluster correlation coefficients of two joined clusters m and l;
2) the minimum level r_{ij} should not be lower then the threshold value r_{cr} and the average correlation coefficient $\bar{r}_{l+m}$ should not to be too low in comparison with the major part of the initial clusters;
3) only clusters that are geographically close should be integrated.

The integration of clusters is based on the use of matrices $\left\|\bar{r}_{m,l}\right\|$ and $\left\|r_{min,m,l}\right\|$, where m and l are the number of the clusters. Sometimes the matrix $\left\|r_{max,m,l}\right\|$ can also be of use.

Analysis of single stations

All stations (not integrated in the clusters) are divided into peripheral (situated at the fringe of a region) and central ones. Peripheral stations can usually not be joined to the clusters without violation of the requirement $r_{min.} \geq 0.2$. Part of the central stations can be joined to the clusters using the criteria for joining clusters.

Final representation of data

A map of the regions sharing the boundaries of clusters (Figure 1) illustrates the common picture of synchronicity of annual runoff fluctuations over the considered territory. The way boundaries between clusters are drawn on a map depends on the density and uniformity of the spatial disposition of observation stations. Mountain territories with clearly expressed orographic barriers can be accepted. In flat territories more or less wide bands separate detected clusters. Therefore the preliminary character of the boundaries is inevitable and, apparently, is sufficient for the majority of tasks studying synchronicity of river runoff fluctuations.

The characteristics of the derived regions are shown in Table 3, where the main diagonals of the matrices of inter-cluster correlation $\left\| \bar{r}_{k,l} \right\|$, $\left\| r_{min,k,l} \right\|$, $\left\| r_{max,k,l} \right\|$, $\left\| d_{k,l}(r > 0.5) \right\|$, $\left\| N_{k,l} \right\|$, $\left\| N_{k,l}(r > 0.5) \right\|$ are represented. . The following abbreviations are used in the table: $N_{k,l}$ - common number of correlation coefficients between objects of regions k and l, $N_{k,l}(r>0.5)$- number of correlation coefficients $r_{ij}>0.5$ and $d_{k,l}(r>0.5)= N_{k,l}(r>0.5)/ N_{k,l}$.

The structure of an inter-regional correlation is characterized by a set of extended matrices by adding other thresholds of R - $\left\| N_{k,l}(r > R) \right\|$ and $\left\| d_{k,l}(r > R) \right\|$ (Table 4).

TABLE 3. Characteristics of regions with synchronous river runoff fluctuation (Figure 1)

Region	I	II	III	IV	V	VI	VII	VIII
N	11	7	9	8	11	15	16	7
N_r	55	21	36	28	55	105	120	21
$\bar{r}$	0.564	0.572	0.569	0.599	0.624	0.605	0.574	0.714
r_{min}	0.250	0.275	0.265	0.338	0.333	0.219	0.200	0.397
r_{max}	0.912	0.907	0.869	0.854	0.906	0.925	0.910	0.940
$N_{(r>0.5)}$	36	15	23	20	43	72	81	20
$d_{(r>0.5)}$, %	65	71	64	71	78	69	68	95

TABLE 4. Characteristics of regions with synchronous runoff
Mean of correlation coefficients (upper part) and number of correlation coefficients (lower part)

Region	1	2	3	4	5	6	7	8
1		0.273	0.034	-0.121	0.243	-0.054	-0.047	0.198
2	77		0.075	0.092	0.166	-0.003	0.264	0.4
3	99	63		0.254	0.416	0.273	0.131	0.19
4	88	56	72		0.117	0.315	0.14	0.108
5	121	77	99	88		0.19	0.003	0.441
6	165	105	135	120	165		0.202	0.116
7	176	112	144	128	176	240		0.134
8	77	49	63	56	77	105	112	

Minimum (upper) and maximum (lower) of correlation coefficients

Region	1	2	3	4	5	6	7	8
1		-0.087	-0.29	-0.453	-0.245	-0.396	-0.43	-0.296
2	0.792		-0.244	-0.326	-0.129	-0.457	-0.115	0.138
3	0.349	0.345		-0.088	0.024	0.061	-0.133	-0.108
4	0.253	0.477	0.646		-0.171	-0.271	-0.158	-0.158
5	0.811	0.417	0.853	0.579		-0.127	-0.397	0.078
6	0.304	0.36	0.618	0.749	0.529		-0.241	-0.262
7	0.274	0.606	0.381	0.404	0.319	0.556		-0.179
8	0.643	0.675	0.446	0.499	0.725	0.404	0.713	

Ratios N(r< - 0.2)/Nall % (upper) and N(r>0.2)/Nall % (lower)

Region	1	2	3	4	5	6	7	8
1		0	2	32	2	9	15	6
2	68		3	13	0	4	0	0
3	14	17		0	0	0	0	0
4	1	38	61		0	1	0	0
5	54	39	92	24		0	6	0
6	2	5	70	72	39		1	1
7	3	63	30	33	4	55		0
8	57	94	44	25	99	28	28	

Number of coefficients R< - 0.2 (upper) and number of coefficients R>0.2 (lower)

Region	1	2	3	4	5	6	7	8
1		0	2	28	2	15	27	5
2	52		2	7	0	4	0	0
3	14	11		0	0	0	0	0
4	1	21	44		0	1	0	0
5	65	30	91	21		0	11	0
6	4	5	94	86	65		2	1
7	5	71	43	42	7	131		0
8	44	46	28	14	76	29	31	

Ratios N(r> 0.5)/Nall %(upper) and N(-0.2<r<0.2)/Nall (lower)

Region	1	2	3	4	5	6	7	8
1		6	0	0	10	0	0	6
2	32		0	0	0	0	10	27
3	84	79		6	35	4	0	0
4	67	50	39		1	19	0	0
5	45	61	8	76		1	0	35
6	88	91	30	28	61		3	0
7	82	37	70	67	90	45		5
8	36	6	56	75	1	71	72	

4. A case study for the territory of the former USSR (except Central Asia)

This approach will be illustrated on a second case study of regionalisation of the territory of the former USSR (except Central Asia). The data set consists of 227 annual discharge series for 1965-85 having an average observation period of 20 years. Regionalisation is carried out using the same procedure described above for the East European Plain. The resulting regions are shown in Figure 2.

Although the focus of attention here is on methodological aspects it is worth noting the modifications in configuration of regions on the East European Plain for this map compared with the one shown in Figure 1. These modifications are basically explained by the change of spatial runoff correlation caused by the significant alteration in the character of the atmosphere circulation in the first half of the 1970s. During those years the domination of the alternating forms of meridional circulation was replaced by a phase of growth of the zone of circulation recurrence [4].

As a first approximation it is reasonable to analyze synchronicity of runoff fluctuation with the help of the matrices $\left\| \bar{r}_{k,l} \right\|$ and the graphs of Figure 3, giving insight into the structure of inter-cluster correlation and facilitating a visual analysis of data. For example, if the mean correlation between clusters k and l is $\bar{r}_{k,l}$ = 0.3 then it can be approximately established from the graphs that $d_{k,l}(r > 0.2) \approx 70\%$, $d_{k,l}(r > 0.5) \approx 15\%$, $d_{k,l}(r<-0.2)\approx0\%$, $r_{max.k,l} \approx 0.7$ and $r_{min.k,l}\approx0.1$. On the other hand, if the criterion $\bar{r}_{k,l} < -0.25$ is used as a diagnostic for non-synchronicity, we obtain the following structure of correlation: $r_{max.k,l}\approx0$, $r_{min,k,l}$ between -0.6 to -0.7, $d_{k,l}(r>0.2)\approx 0$, $d_{k,l}(r<-0.2)\geq60\%$ and $d_{k,l}(r<-0.5)\geq 10\%$. In this latter case, practically all estimates of r_{ij} between the rivers of region k and rivers of region l are negative and the major part of them is significantly less than zero. A considerable part is composed of negative coefficients with absolute values in the interval 0.5 to 0.7. Similar types of information allows justification, or at least explanation, in wording the description of geographical regularities in the synchronicity of runoff fluctuations. It is of special importance when using terms designating the degree of synchronicity (non-synchronicity) like "weak" or "strong".

5. Conclusions

The identification of regions of synchronic runoff patterns presented here is one of several possible approaches relying on spatial runoff correlations. Information loss and subjective judgments are inevitable whatever method is used. The proposed approach is aimed at making a reasonable minimization of these negative factors.

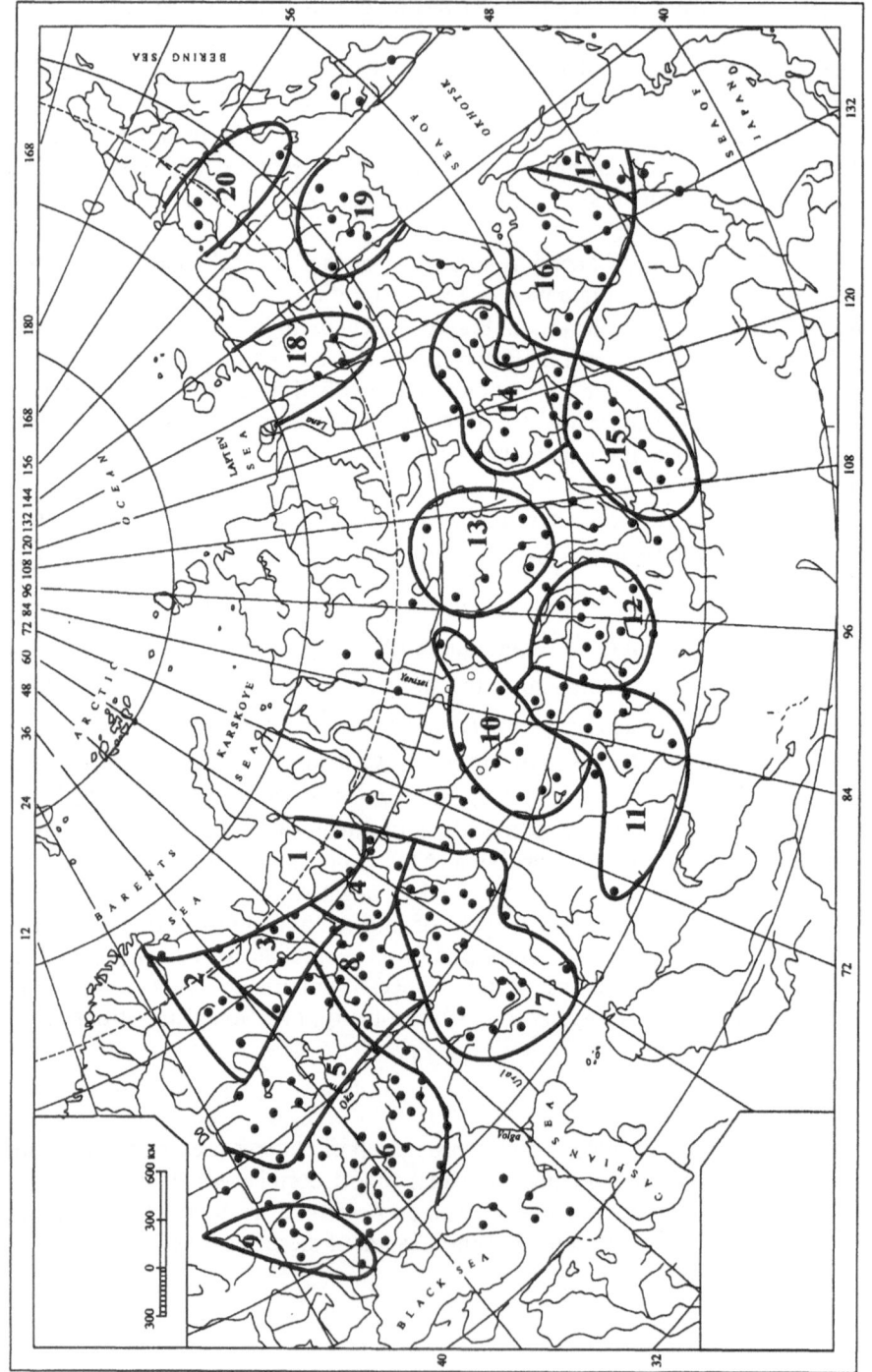

Figure 2. Regionalization of the former USSR territory for synchronizm of annual runoff fluctuations (1965-1985)

Legend: • basin centers, **10** - regions

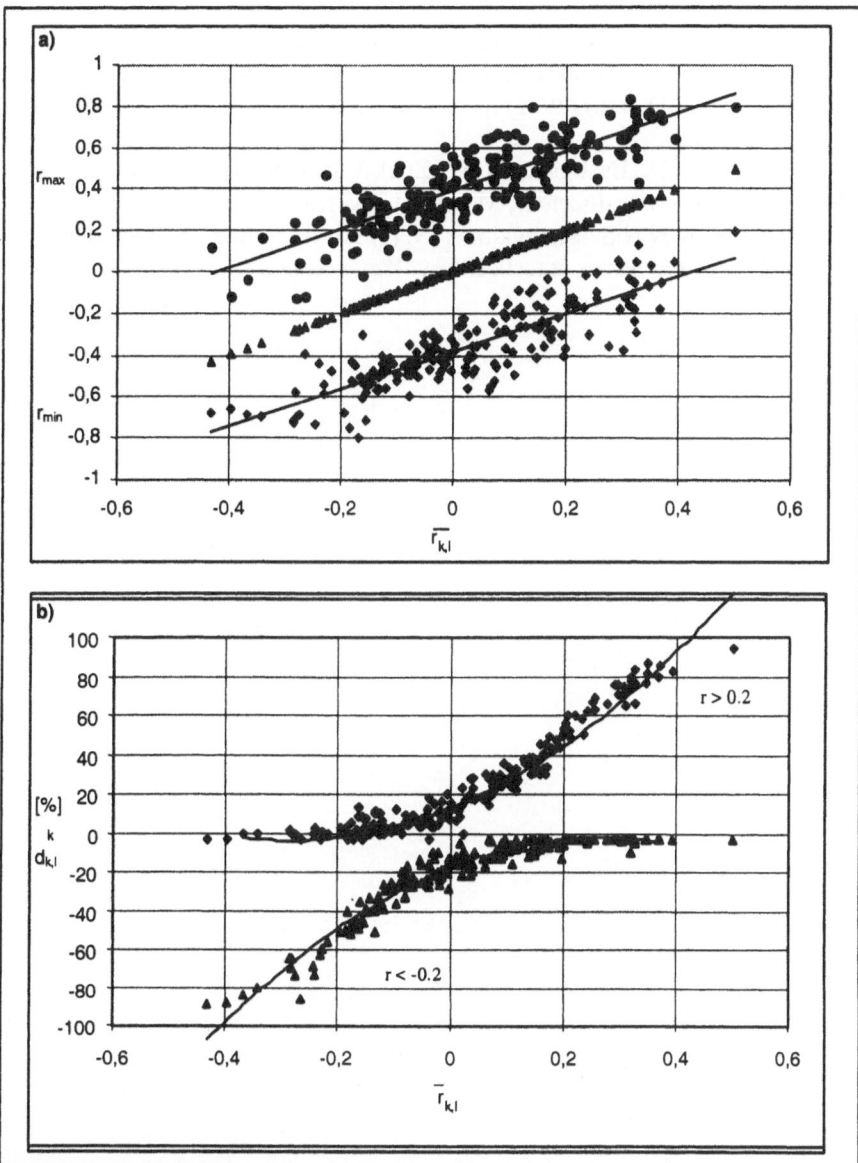

Figure 3. Relationships between synchronicity (non-synchronicity) characteristics and mean of interregional correlation coefficients:

a) maximal and minimal correlation coefficients between objects of different regions;

b) quotas of estimations $r_{ij}<-0.2$ and $r_{ij}>0.2$ in common number of correlation coefficients

6. References

1.Evstigneev V.M.(1985) Methods of pattern recognition and study of synchronicity of runoff fluctuations. *Vestnik of Moscow University*, Geography, №1, 10-15.

2. Bykov V.D. (ed.) (1982) *Research and calculation of runoff*, Moscow University Publishers.

3. Bykov V.D., Evstigneev V.M., Zhuk V.A. (eds.) (1984) *Calculation of runoff (Methods of spatial generalization)*, Moscow University Publishers.

4. Sidorenkov N.S., Shveykina V.N. (1996) Alteration of climatic behavior of Volga and Caspian sea basins at last century. *Water Resources*, Vol.23, №4, 401-406.

STOCHASTIC MODELS WITH PERIODIC-CORRELATION OF SEASONAL RIVER RUNOFF VARIATIONS

M.V. BOLGOV
Water Problems Institute, Russian Academy of Sciences
117735, GSP-1, Gubkin St., 3, IWP RAN, Moscow, RUSSIA

1. Introduction

Development of stochastic models for seasonally varying processes is one of the problems insufficiently studied in hydrology. The complexity of this problem arises in the nonstationary character of these processes (or, to be more specific, their periodic correlation), the lack of adequate mathematical tools, the necessity for consideration of a large number of parameters, and the computational instability of the methods used. Specialists in this field suggest different approaches varying in complexity for modelling hydrological processes involving seasonal variations. The method of fragments and the method of sequential determination of linear autoregression, suggested by G.G.Svanidze[14]; the method of canonical expansions, suggested by Busalaev and Davletgaliev [3]; the method of normalisation, modified by Reznikovsky and others [10], are some examples of these methods. A large number of approaches for solving this problem are touched upon in [3,13,16]. Priestley [6] and Tong [16] recommend complex non-linear models. The above methods reproduce with different certainty the properties of the observed series but have a common limitation. When studying the seasonal runoff variations, the performance in terms of two-dimensional and multi-dimensional distributions by the empirical data is not evaluated, and the properties of the stochastic models, as well as their optimal complexity are usually not discussed.

2. Seasonal Runoff Variations

The processes involving seasonal runoff variations are obviously nonstationary, and the mathematical models used for their description are very complicated and cumbersome. In addition, it is necessary to note that a relatively strict mathematical theory has been developed only for the processes with the so-called "correct" seasonal variations [7,11], and it cannot be directly extended to the "irregular" series of monthly runoff values.

The complexity of the stochastic nature of runoff has led to numerous simplified solutions, with different degrees of details in the properties of the real processes and their number. Some versions of a periodic autoregression model will

M.V. Bolgov et al. (eds.), Hydrological Models for Environmental Management, 51–66.

be discussed below. Emphasis is put on autoregression models because of the relatively simple calculation of the effective estimates of their parameters.

In the case of stochastic modelling of river runoff for a time interval of less than a year, it is advisable to consider the river runoff as a nonstationary random process with parameters whose variation period is equal to the year. The analyses of nonstationary time series involve: (a) normalising one-dimensional stochastic models of runoff; (b) selecting the type of stochastic model of the process; (c) selecting the optimal complexity of the model; (d) selecting the method for modelling; (e) assessing the degree of agreement of the model properties to the observations.

Let us proceed to a detailed discussion of these problems. The first and very important problem is normalising the runoff distributions as the theory of autoregression has been developed mainly for the Gaussian case, while the distributions of the mean monthly runoff are characterised by higher skewnesses reasonably well approximated by the three-parameter gamma distribution (Table 1).

Table 1. Parameters of the distribution of the mean monthly runoff, the Unzha River in Makaryev

Month	Mean runoff, m^3/s	Coefficient of variation	Coefficient of skewness
Jan.	44.3	0.49	4.0
Feb.	36.5	0.38	5.0
Mar.	37.6	0.34	5.0
Apr.	72	0.61	2.0
May	691	0.47	1.0
June	175	0.61	2.0
July	110	0.82	4.0
Aug.	74.0	0.81	4.0
Sept.	90.4	1.15	4.0
Oct.	123	0.90	4.0
Nov	118	0.71	2.5
Dec.	72.0	0.71	3.5

A transformation of the normally distributed values to the gamma distributed can be carried out in several ways. A suitable way, recommended earlier [e.g. 4,10], is a transformation of the nonstationary gamma process (river runoff) into a series of uniform distributed random processes in [0,1] and then into the Gaussian process with a zero average and unit variance. Assuming that the Gaussian series are described by a stochastic model, which is linear with respect to the regression equation, the two-dimensional densities of the gamma values needed for estimating the ordinates of the correlation functions, may be obtained by transforming the classical two-dimensional normal distribution [5]. However, it should be noted that the gamma correlation (two-dimensional density) in our case is asymmetrical.

Having transformed the gamma distributed runoff values into the series of normally distributed values with a zero average and unit variance, a stochastic model can be constructed. One of the most widespread methods is the withdrawal of the seasonal variations from the observed series and subsequent normalisation. The resulting series is considered as stationary and is presented as a common autoregression model:

$$Q_t = \varphi_1 Q_{t-1} + \ldots + \varphi_p Q_{t-p} + \varepsilon \tag{1}$$

where ε are independent and equally distributed Gaussian random values.

The estimates of the parameters of the model eq. (1) may be found using one of the known procedures [1]. In this case, we consider an autoregression model for seasonal variations, or the *SAR* model, so called by Privalsky et al. [7].

The difficulty when using the *SAR* model consists in fitting an autoregression model for each month. Having estimated the autoregression parameters and white noise variance for each month, we obtain a periodic autoregression model (a *PAR* model). The agreement of these models with the available observations is then considered. The following tests of the adequacy of the model are suggested:
(1) sample autocorrelation functions of the observed and simulated series must be in agreement within a confidence interval;
(2) the autocorrelation of the mean annual runoff values must be preserved in the simulated series.

The runoff observations for the Volga River (in Volgograd), the Unzha River (in Makaryev), and the Kolyma River (in Srednekolymsk) have been used in this study as examples. All these rivers are mainly snow-fed but rain contribution is significant for the Kolyma River. The discretization interval here is one month or less.

Autocorrelation functions for monthly runoff differ significantly for different months and for different seasons in particular. That is why an application of the SAR model in this case is not possible as, after the withdrawal of seasonal variations and normalisation, the obtained series have a correlation structure depending on the number of the calendar month, i.e. the periodical correlation of the process is preserved. Thus, periodic autoregression models of different orders remain at disposal for the analysis. Each month in the autoregression model is assigned a number. This means that the regression coefficients and white noise variance will depend on the number of the month.

An autoregression model for each month can be obtained by solving the Yule-Walker system of equations and estimating the white noise variance. It is important to assure that the stability of the autocorrelation is preserved even for low orders. In order to improve the stability of the estimates obtained with the help of the Yule-Walker system of equations, it is recommended to smooth the sample autocorrelation functions. A more smooth autocorrelation function estimate obtained using a more complicated model, should be substituted in the Yule-Walker system of equations.

The next problem is selecting the order of an autoregression model. Svanidze [14] recommends considering the correlation with eleven preceding values. Reznikovsky et. al. [10] advise not to use models of a higher order than three. The ensemble of autocorrelation functions shows that it is necessary to assign a certain order of model complexity for each month.

In cases of significant annual autocorrelation of the runoff for the adjacent years (this is typical for most of the area of the former USSR), the complexity of the model for individual months must be assigned striving at preserving a certain correlation coefficient between runoff of the adjacent years. Otherwise, the possibility of practical application of the recommended stochastic model sharply decreases. Table 2 presents the orders of the autoregression models for each month for the three rivers studied.

Table 2. The orders of the autoregression models for calendar months

Month	The Kolyma River	The Volga River	The Unzha River
Jan.	1	11	11
Feb.	2	11	11
Mar.	2	11	11
Apr.	3	1	2
May	1	2	2
June	1	11	11
July	2	11	11
Aug.	2	11	11
Sept.	1	11	11
Oct.	2	11	11
Nov	2	11	11
Dec.	2	11	11

It is seen that the order of models increases for periods with a dominant stable ground-water contribution to the runoff formation. The order decreases to one or two for the months with floods. Having selected the order of the model and calculated white noise variances, we can proceed to autoregression coefficients, which are used for modelling a periodically correlated nonstationary process. For this purpose, it is necessary to simulate a normally distributed random value with a dispersion equal to the white noise variance, and using equation (1) simulate the values for the process one-step ahead. The same procedures are then repeated a required number of times for all the models. Here, it should be noted that the PAR model consisting of a population of twelve "quasi stationary" autoregression models is based on an assumption that should be properly investigated. To reveal the extent of the introduced distortions, let as simulate an artificial long sample using the mentioned scheme and evaluate its parameters. In this study, monthly runoff values for 2000 years for the three rivers investigated have been simulated.

As expected, in the case of fairly simple models (of the order one to three), the method for constructing the PAR model does not introduce significant deviations. However, when the complexity of the model increases, a positive shift is observed when estimating the variance of the modelled series (table 3). The results in Table 3 show that the Gaussian distribution holds because the sample skewness estimates are close to zero and the estimates of the fourth central moment (excess) are about 3.

The Gaussian character of the distribution makes possible an introduction of a correction for shift by means of normalisation by the value of a sample mean square deviation. As a linear transformation does not change the correlation coefficient, the form of the autocorrelation function will be preserved in the normalised series. The introduced distortions have the following reason. Runoff modelling for each month is done using the model for the given month and runoff values, simulated at the previous steps with the help of other models not adequate for the stochastic scheme of the given time step. The results of the simulations show that the series of monthly runoff is reproduced with smaller distortions for a smaller order of the autoregression model. Distortions are totally absent in the Markovian case.

Table 3. Parameters of the distributions of the periodically correlated Gaussian random process, generated with the help of the PAR model for the Volga River.

Month	Average	Mean square deviation	Skewness	Excess
Jan.	0.014	1.190	0.152	3.01
Feb.	0.0	1.291	0.141	2.96
March	0.002	1.270	0.091	2.92
April	0.007	1.028	-0.034	2.84
May	-0.009	1.017	0.063	2.81
June	0.021	1.019	0.046	2.96
July	0.030	1.031	-0.037	3.10
Aug.	0.018	1.067	-0.005	3.05
Sep.	0.015	1.082	-0.007	2.96
Oct.	0.026	1.110	0.083	2.91
Nov.	-0.001	1.147	0.072	2.84
Dec.	-0.008	1.172	0.177	3.06

The quality of the simulated series is assessed with the help of another hydrological test considering the correlation of annual runoff values. As seen from Table 4, the agreement between the values of the correlation coefficients for the runoff values for adjacent years for the observed and estimated series is satisfactory.

Table 4. Coefficient of correlation for the runoff values for adjacent years.

r(1)	The Kolyma River	The Unzha River	The Volga River
Observed series	0.0	0.3	0.4
Simulated series	0.0	0.2	0.3

3. Empirical estimates of the parameters

A stochastic runoff model capable of reproducing the properties of the observed series both with respect to the long-term and the seasonal runoff variation has not yet been created. The achievements in this field differ in ideas and methods for different levels of discretization. Certain versions of the Markovian models with one-dimensional gamma distributions are suitable for annual runoff series. Some authors [7,14] recommend various versions of autoregression model for mean monthly runoff series. These latter models can be much more complicated. The use of stochastic models for solution of application problems in engineering hydrology is caused, as mentioned above, by the need in methods and models to reproduce the properties of the observed series for the whole ensemble of time intervals. The problem is that the parameter estimates, determined from limited observational data, fail to reproduce the properties of the process at the level of the mean annual values when analysing the seasonal variations. In these cases, "rough" estimations sometimes yield good approximations, but they are not theoretically founded.

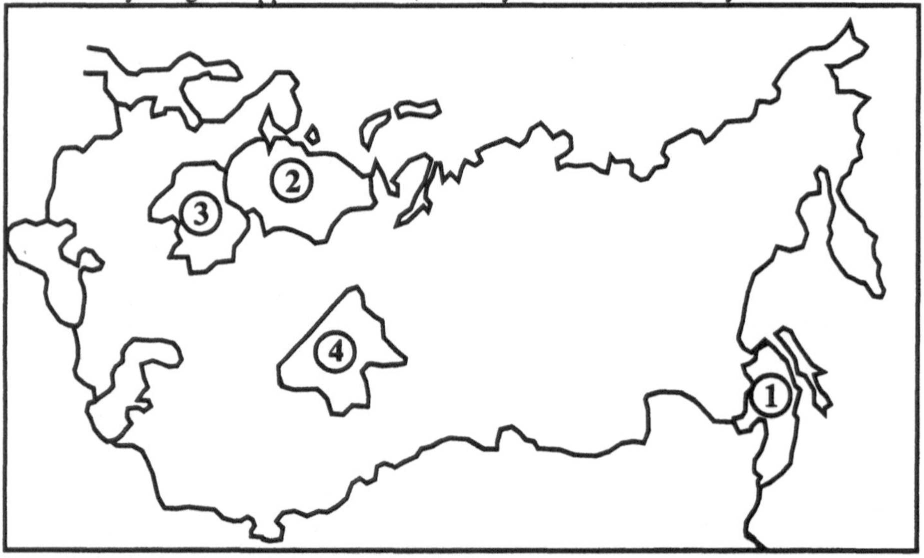

Figure 1. Location of the areas studied (number of region in accordance with Table 7).

The defined problem governs a further analysis of the observation data. As noted in numerous works on long-term runoff variations [4,8,10,14], there is a positive correlation between adjacent sample terms $r(1)$ in the mean annual runoff series. Different hypotheses were proposed to explain the causes of autocorrelation. In [2] an attempt was made to obtain area average parameters of intraseries autocorrelation (within statistically and genetically homogeneous areas). The results obtained in [2] were used to select four hydrological areas for studying properties of the interannual runoff variation (Fig.1). These areas have significantly different autocorrelation coefficients of annual runoff series ranging from $r(1)=0.06$ in the Primorye region (area 1) to $r(1)=0.37$ in the Irtysh and Ishym basins (area 4).

The area boundaries might change in the future as the study continues, but this subdivision is useful anyway to determine the most general statistical properties and regularities. On the one hand, the dimensions of the areas allow one to ignore asynchronous runoff variations caused by geographical zonality. On the other hand, it is possible to average the sample estimates over a large number of observation sites (about 100). Only stationary (in terms of the conditions of runoff formation) and at least 20 year long series were studied statistically. Rivers stretching over several geographical zones were also omitted. The rivers of three out of the four areas studied (2, 3, and 4) have a spring-flood. The rivers of the Primorye region have floods during the warm season (June - September). The results of the generalization of the sample estimates of the ordinates of the autocorrelation functions are shown in Fig.2 as three-dimensional surfaces and isolines. Note that the averaged estimates were obtained for the normalized (not to be confused with standardized) runoff sequences. The normalization procedure involved the following two steps:

- transformation of the initial series with a three parameter univariate gamma distribution into the series of uniformly distributed numbers in the [1,0] interval;
- transformation of the uniformly distributed variables in to normally distributed variables with zero mean and unit variance.

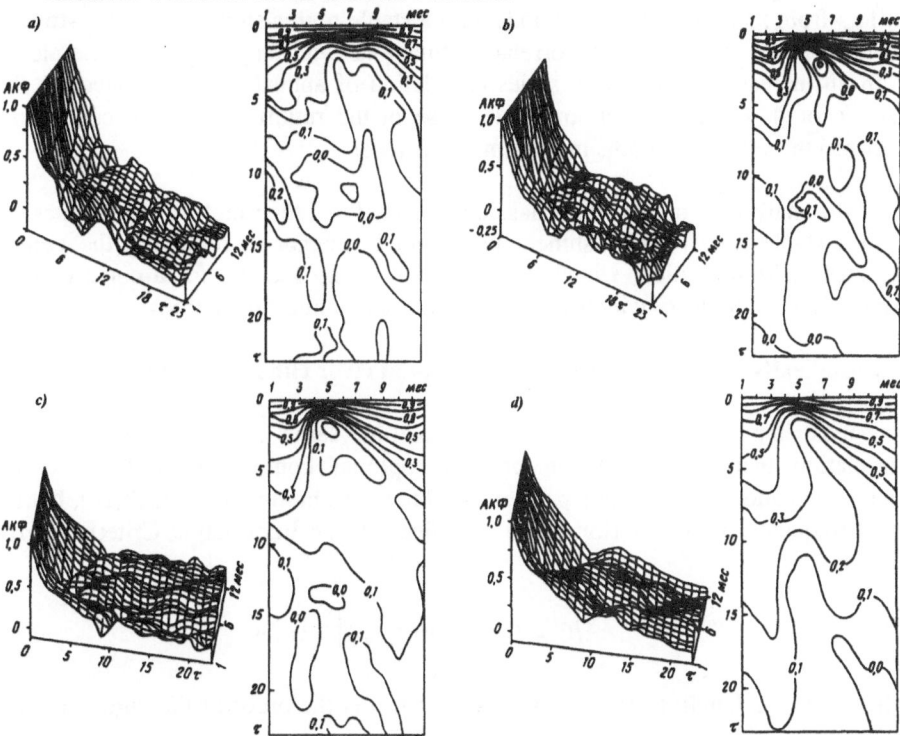

Figure 2. Averaged sample estimates of Autocorrelation function (*ACF*) ordinates shown as three-dimensional surfaces and isolines depending on the month number for areas *1(a)*, *2(b)*, *3(c)*, and *4(d)*.

As can be seen in Fig.2, the correlation structure of the normalised series depends on the number of the month in all areas due to great differences in runoff genesis during different seasons. Stable low-water periods are characterised by high runoff correlation coefficients of adjacent months and weak attenuation of the autocorrelation function as a whole, which can be readily explained by the dominance of the groundwater component in river replenishment. The runoff during spring (high water) months depends weakly on the preceding runoff. Autocorrelation functions for high-water months are decreasing quickly. The autocorrelation function for rivers with the prevailing rainfall replenishment during the warm season (Primorye region) decrease quickly practically for all the months during the high-water period. The autocorrelation functions for the months following a flood period become less steep and the intraseries relations increases.

The autocorrelation functions for low-water periods in all areas are characterised by a minimum corresponding to the correlation of low-water with a preceding flood. The generalised estimates of the monthly runoff autocorrelation functions with large time lags (more than one year) differ strongly for different areas depending on the runoff autocorrelation coefficient in adjacent years. The following regularity could be seen. When the correlation between the runoff of the adjacent years is negligibly low (Primorye region), the autocorrelation function scarcely differs from zero even during periods with a stable replenishment for a lag -time of more than one year. The autocorrelation functions with large lags decrease slower with an increase in the within series correlation of annual runoff. Neglecting this feature of *ACF*s leads to a strong decrease in the runoff correlation coefficients between the adjacent months in the simulated series.

The analysis of observations has demonstrated that the runoff during the periods with a stable runoff replenishment has a good "memory", reflecting the genetic features of its formation, which should be taken into account for a correct runoff simulation in engineering hydrology and hydro-economic calculations.

4. Complexity of stochastic models of seasonal river runoff variations

The most important stage in the analysis of time series in the framework of the correlation theory is a choice of their optimal complexity (order). A few recommendations to solve this problem are known from the literature [7,10], but the most frequently used criterion is the so-called Akaike Information Criterion *(AIC)* [7,16]:

$$AIC = Ln\big[(n-p)\sigma_\varepsilon^2(p)\big] + 2p/n \qquad (2)$$

where n is the length of the samples analysed, p is the order of the autoregression model and σ_ε^2 is the squared approximation error ("white noise" variance).

The expression for *AIC* consists of two summands. The first one usually decreases with increasing p, while the second increases. The p value, for which the

AIC has the minimum, is considered optimal. The Akaike criterion is usually applied to stationary series, and its application to periodically correlated nonstationary series needs to be properly justified. To determine the optimal complexity of stochastic models for different months of the year within the selected areas, individual values of the order p at which AIC is a minimum, have been identified for each site and each month. The optimal values of the order of the autoregression models (one decimal accuracy) averaged within the areas for each month are presented in Table 5. In most cases the optimal order (complexity) of the model varied between two and three.

The results in Table 5 show that the Akaike criterion generally yields a model order that is much lower than the expected one. The earlier studies have shown that such an order of model is suited only for rivers with no runoff autocorrelation between adjacent years. If the autocorrelation between annual runoff values is positive, then the complexity of the model should be increased. The demand for preserving the autocorrelation of annual mean runoff values in the simulated series may serve as a unique criterion for the choice of the complexity of the corresponding models.

It is necessary to stress that the observational series are short. The sample variance of the model parameters determined from a short sample can be fairly large. The criterion becomes less sensitive. The following simulation experiment was performed to determine the degree of influence of the sample size on the order of the autoregression model (sensitivity of the criterion). Runoff observations for the Volga River near Volgograd (for 60 years before construction of the Volga reservoirs) were used to determine parameters of 12 autoregression models for each month. The order of the model was assumed to be one for April, two for May, and eleven for the other months. The set of 12 autoregression models was used to simulate the artificial 10000-year long monthly runoff series. Here we will limit ourselves to the normally distributed numbers with a zero mean and unit variance, i.e. normalised runoff values.

Table 5 The optimal values of the order of the autoregression models

Month	The Upper Volga River	Primorye region
Jan.	2.3	2.1
Feb.	2.3	2.2
Mar.	2.5	1.9
Apr.	1.8	2.0
May	1.5	2.0
June	2.2	1.5
July	2.0	2.0
Aug.	2.2	1.5
Sept.	2.2	1.9
Oct.	2.1	2.5
Nov	2.0	2.4
Dec.	2.4	2.0

Table 6. The mean values of the orders of autoregression models determined on the basis of short N years long samples for the simulated series with seasonal variations

Month	Sample size N			
	25	50	100	200
Jan.	1.9	2.2	2.1	2.4
Feb.	1.9	2.3	2.6	3.6
March	2.1	2.0	2.4	2.9
April	1.8	2.9	3.9	4.5
May	2.0	2.5	2.8	3.5
June	1.8	2.2	2.4	3.0
July	1.9	2.5	3.1	4.0
Aug.	1.9	2.1	2.8	3.4
Sep.	1.9	2.5	3.2	3.8
Oct.	1.9	2.1	2.3	3.1
Nov.	1.9	2.0	2.2	2.3
Dec.	1.8	2.1	2.2	2.6

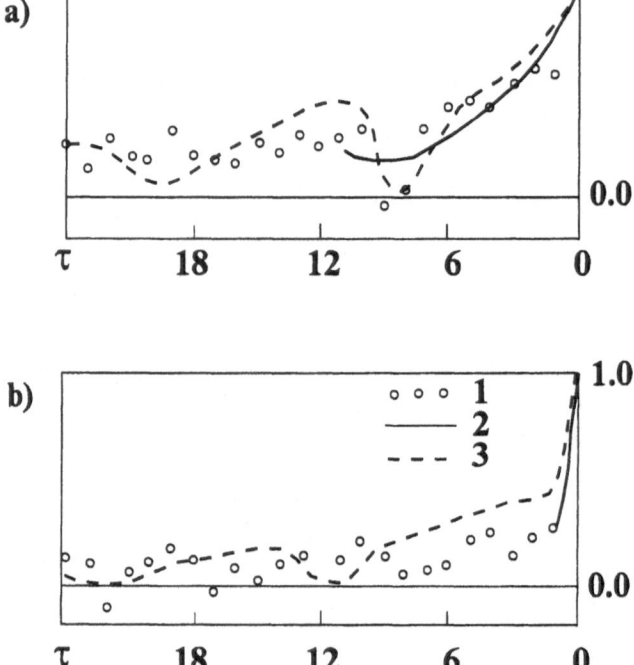

Figure 3. Autocorrelation functions of mean monthly discharge of the Volga River for January *(a)* and April *(b)*. **1** - *ACF* estimates from the observation series; **2** - ACF smoothed estimates; **3** - *ACF* estimates for the simulated artificial series for N=2000 years (*PAR*-model).

The series obtained were subdivided into the set of short samples of 25, 50, 100 and 200 years. The optimal order of the model for each month was determined for all the short samples, and the values obtained were then averaged. Table 6 presents the average values of the orders of the autoregression models for the short series of different length (one decimal accuracy). It can be concluded from this table that the observation period available at present (i.e. 50-100 years) is not suitable for determining the order of the required stochastic model using the Akaike criterion. If the series length is increased to 200 years, the criterion indicates a higher order model, but this increase in the order is insignificant in comparison to the order of the initial model. It is interesting that when the sample size is large, the Akaike criterion points at higher order models for the months that have been simulated using the simplest models. To illustrate this effect, Fig. 3 shows the autocorrelation functions for January and April, both originally used in the PAR model and simulated in the artificial series. For January, the smoothed autocorrelation function used for modelling is in good agreement with the empirical data and, in spite of the high order (11th), is reproduced by the model fairly well. The situation in April is quite different. A spring flood that begins during this month indicates a significant change of the replenishment source. However, the assumption that the April runoff is weakly correlated with the runoff of the preceding months yields a more intricate form of the autocorrelation function for this month for the modelled series with small lags, which is indicated by AIC. This effect is evidently caused by the inconsistency between the stochastic models for individual months which use smoothed estimates of the autocorrelation functions to diminish the instability in the solution of the Yule-Walker equation system with large order of magnitude.

Discussion of the modelling methods is continued on a real example, namely, by using the autocorrelation functions of the Volga River runoff near Volgograd calculated for a period of over 60 years. Early consideration was given to a modelling method based on the *PAR* model and smoothing the autocorrelation functions for each month. The method revealed a positive bias of the variance of the modelled series but, nevertheless, yielded a satisfactory agreement between the normalised autocorrelation functions of the observed and modelled series. The reasons for a significant bias of the variance have already been examined earlier, but let us consider some other modelling methods including the traditional method of fragments.

According to Svanidze [14], a double-sample method (or method of fragments) reduces the modelling of the series with a seasonal variation to two steps: 1) modelling the annual runoff series according to a given stochastic scheme and 2) selecting the realisation of the intra-annual distribution at random from the available set of fragments (set of observed realisation of seasonal variation). The autocorrelation function in Fig.4, calculated on the basis of the 2000-year series modelled by a double-sample method, for some months agrees fairly well with the empirical estimates. The autocorrelation functions decrease slowly, which is a result of the adopted modelling method that permits conservation of the runoff correlation between the adjacent years in the artificial series. At the same time, Fig. 4 shows a discrepancy between the autocorrelation functions found from the modelled series and empirical estimates calculated from observed data for the months with stable replenishment. The autocorrelation functions fail to reproduce the minimum, which reflects the weak correlation with the runoff of the previous flood resulting from the limited number of fragments (in this case 60 realisations of the seasonal pattern).

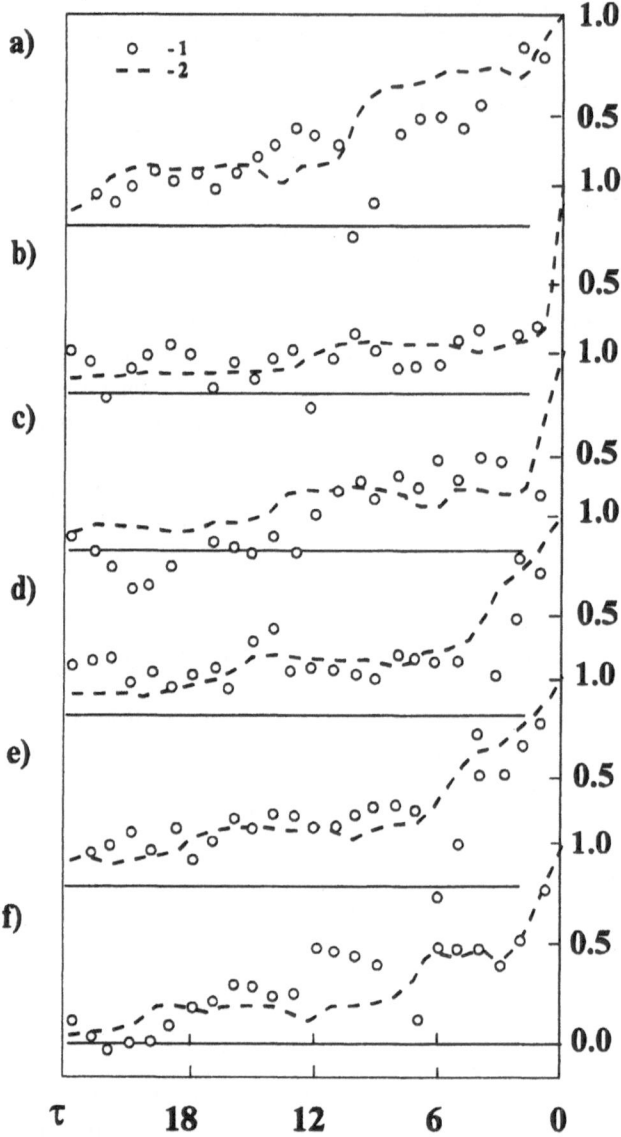

Figure 4. Autocorrelation functions for the modelled series (method of fragments): *a* - February, *b* - April, *c* - June, *d* - August, *e* - October, *f* - December. 1 - *ACF* estimates from the observation series; 2 - for the modelled series.

Applying the *POLAR* method (sequential determination of linear autoregression) [14], Svanidze also proposed using the correlation of monthly runoff not only with the runoff for the previous month but also with mean annual runoff for both the previous two years and the year for which the seasonal pattern is reproduced. Note that in the autoregression scheme it is not necessary to consider the correlation of the runoff for a month with the runoff for the same year, since this complicates the model as the autoregression coefficients describing the runoff of each month take this correlation into account.

Consider the following modelling scheme with the same empirical data for the Volga River. Third -order autoregression models will be used for all low-water months and, as before, first and second - order models for the high-water months. Additionally, we will introduce to the autoregression scheme the correlation of the runoff of the given month with the mean annual runoff of the previous year. For the Volga these correlation coefficients for months from December to January are respectively: 0.69, 0.55, 0.67, 0.17, 0.25, 0.26, 0.21, 0.37, 0.42, 0,39, 0,34 and 0.23. The 2000-year long series simulated using this scheme is characterised by an autocorrelation function presented in Fig.5. On the whole, there is a satisfactory agreement between the empirical and theoretical autocorrelation functions in the initial sections, i.e. when the time lags are small. For larger lags the autocorrelation functions of the monthly runoff hardly differ from zero, but the modelled series retains the correlation of the runoff between adjacent years.

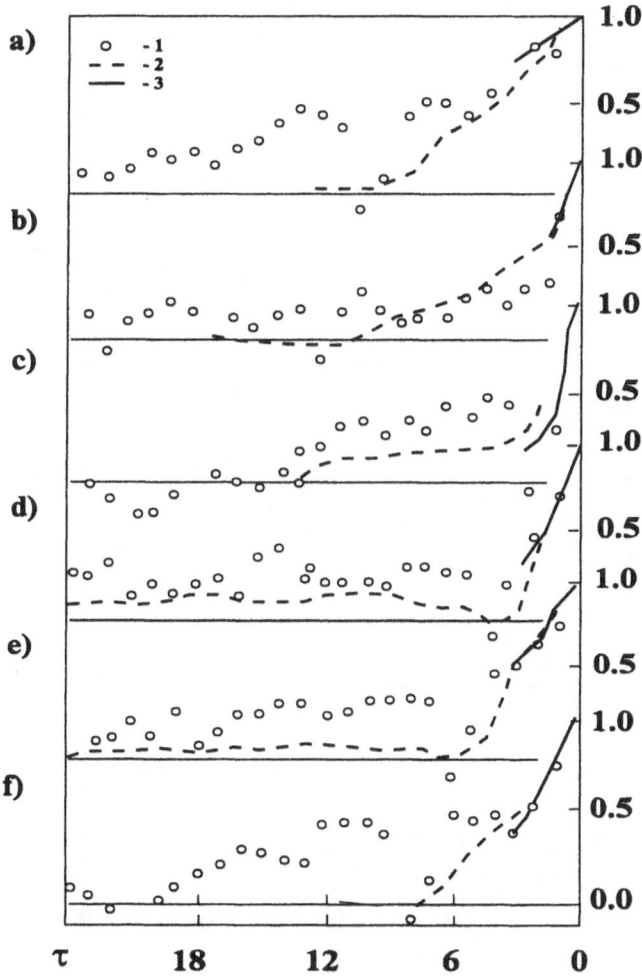

Figure 5. *ACF* for the combined model: 1 - *ACF* estimates from the observation series; 2 - ACF- for the modelled series and 3 - used in modelling.

The advantage of the scheme is that it uses fewer parameters that must be determined (and, hence, specified in the modelling) *viz.* instead of 11 autoregression coefficients for each low-water month, except for the parameters of unconditional distributions, the scheme uses only four parameters. Nevertheless, this approach is explicitly empirical, which on the whole complicates a hydrological analysis and a search for the genetic differences in the runoff formation to explain the differences in the model behaviour in different regions.

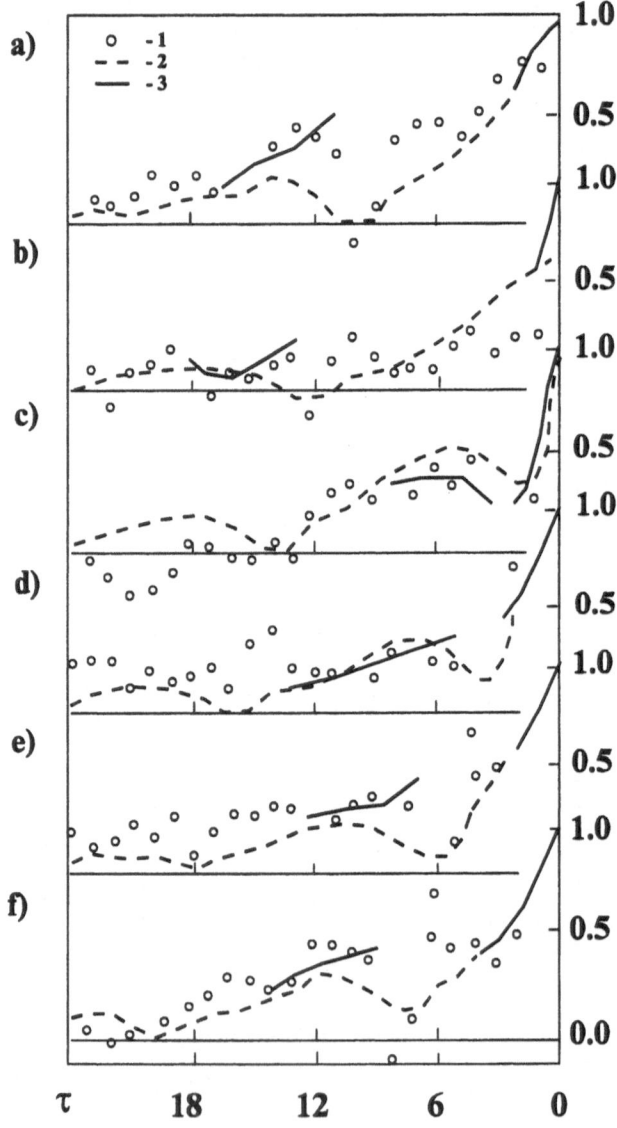

Figure 6. *ACF* for the truncation model: 1 - *ACF* estimates from the observation series; 2.ACF for the modelled series, and 3 - ACF used in modelling.

Let us consider another scheme for modelling the seasonal runoff variations. As before, the autoregression scheme takes into account initial segments of the autocorrelation functions and also the correlation with the runoff of several previous months, except for the high-water period preceeding the given month (Fig.6). The simulation based on this scheme has shown that the model reproduces the autocorrelation functions very close to those for the empirical data, retains the runoff correlation between adjacent years, and significantly (almost threefold) decreases the bias of the variance of the normalised runoff series for the modelled series.

5. Autocorrelation functions

A hydrological problem, like the one considered, puts demands on the model used for reproducing the characteristic features of the observed series in the wide range of periods including time scales from one month up to several (at least two) years. The fact that in the majority of cases the significant value of the inter – annual correlation, $r(1)$ must be taken into account, along with a complicated (non-monotonous) behaviour of the correlation function $r(t)$ for a period less than a year, can lead to rather high values of the order p of the $PAR(p)$ model. To prevent this undesirable situation, we propose to abandon the PAR class by adding a proper white noise to the PAR –process $z(t)$ thereby entering a wider class of $PARMA$ models[1,6]:

$$r(t) = (a-1)\delta(\tau) + a\exp(-\alpha\tau)$$ (3)

Table 7. Parameters a and α of ACF (3) for different hydrological regions (see numbers on Fig.1)

Month	1.Primorye region		2.The North of European Russia		3. The Upper Volga		4. The Middle Ob	
	a	α	A	α	a	α	A	α
1	1.00	0.28	1.00	0.31	0.71	0.14	0.83	0.13
2	1.00	0.26	0.88	0.22	0.82	0.16	0.83	0.12
3	0.75	0.24	1.00	0.24	0.85	0.22	0.74	0.10
4	0.51	0.21	1.00	1.21	1.00	1.46	0.40	0.06
5	0.40	0.18	1.00	-	1.00	1.95	0.30	0.07
6	1.00	1.08	1.00	1.87	1.00	0.69	0.30	0.07
7	0.92	0.64	1.00	1.09	0.85	0.50	0.62	0.16
8	1.00	0.89	0.93	0.57	0.80	0.33	0.73	0.16
9	1.00	0.68	1.00	0.53	0.63	0.15	0.83	0.20
10	0.92	0.41	1.00	0.38	0.71	0.15	1.00	0.24
11	1.00	0.32	1.00	0.44	0.86	0.21	1.00	0.21
12	1.00	0.27	1.00	0.35	0.73	0.18	0.89	0.16

This approach seems to be preferable to the alternative ones, which are rather artificial. The model parameters were determined by the least square method. Their values are presented in Table 7 for some regions. It should be mentioned that the use of individual sample correlation functions is unacceptable for two main reasons: computational instability of the Yule-Walker equations in our case and significant errors in sample estimates. Having this in mind, we used the regionally averaged autocorrelation functions *(ACF)*.

6. Conclusions

The experience of constructing a stochastic nonstationary periodically correlated model of runoff, using the so-called *PAR* model, showed that satisfactory results might be obtained for snow-fed rivers.

Nevertheless, the replacement of a nonstationary process by a population of quasistationary models, which are highly complicated, leads to a positive shift in the variance estimates. However, the Gaussian character of the intermediate series remains, and this makes it possible to eliminate the bias by respective normalisation.

The tests applied showed a satisfactory agreement of the parameters and properties of the observed and simulated series.

7. References

1. Box, J. and Jenkins, H. (1970) *Time series analysis, Forecasting and Control.* Holden-Day. McGraw-Hill Book Co., New York and Maidenhead, Eng.
2. Bolgov M.V., Loboda N.S., and Nikolaevich.N.N. (1993) Spatial generalisation of the parameters of intraseries correlation of annual runoff series, *Soviet meteorology and hydrology,*No. 7,p.p. 83-91, [in Russian].
3. Busalaev, I.V., (1980) *Complex water resources systems,* Alma-Ata: Nauka, [in Russian].
4. Kartvelishvili, N.A. (1981) *Stochastic Hydrology,* Gidrometeoizdat, Leningrad, [in Russian]
5. Moran, P.A. (1969) Statistical inference with bivariate gamma-distribution, *Biometrica,* 56(3), p.p. 123-136.
6. Pristley, M.V. (1988) *Non-linear and non-stationary time series analysis,* Academic press, New York.
7. Privalsky, V.E., Panchenko, V.A., and Asarina, E.Yu., (1992) *Models of time series,* Gidrometeoizdat, St. Petersburg. [in Russian]
8. Ratkovich, D.Ya. (1975) *Long-term river runoff variations,* Gidrometeoizdat, Leningrad. [in Russian]
9. Ratkovich, D.Ya. and Bolgov, M.V. (1996) Conditional probability distributions of seasonal variations in the runoff of snow-fed rivers, *Vodnye Resursy,* 23, p.p. 345-352. [in Russian]
10. Reznikovsky, A.Sh., Alexandrovsky A.Yu., Aturin V.V.,Gladkova S.P., M.A., Kostina, S.G. Romanova E.A.,Rubinschtein M.I.(1979) *Hydrological principles of hydropower production,* Energoatomizdat, Moscow. [in Russian]
11. Rozhdestvenskii, A.V. (1977) *Assessing the accuracy of distribution curves,* Gidrometeoizdat, Leningrad. [in Russian]
12. Rozhkov, V.A., and Trapeznikov, Yu.A. (1977) *Probability models for oceanic processes,* Gidrometeoizdat, Leningrad. [in Russian]
13. Salas, J.D., Delleur, J.W., Yevjevich V., and Lane, W.L. (1980) *Applied modelling of hydrological time series,* Book Crafters, Chelsea.
14. Svanidze, G.G. (1977) *Mathematical modelling of hydrological series,* Gidrometeoizdat, Leningrad. [in Russian]
15. Standards and Specifications 2.01.14-83 (1985) *Determining the design hydrological characteristics,* Stroiizdat, Moscow. [in Russian]
16. Tong, H. (1983) Threshold models in non-linear time series analysis, *Lecture notes in statistics,* **21,** Springer-Verlag, New-York.

IDENTIFICATION OF REGIONAL STOCHASTIC MODELS FROM SHORT TIME-SERIES

M. I. FORTUS
Obukhov Institute of Atmospheric Physics, Russian Academy of Sciences
Ryzhevskii per. 3, 109017 Moscow, Russia
M. V. BOLGOV
Institute of Water Problems, Russian Academy of Sciences
3, Gubkin St., 117971 Moscow, Russia

1. Introduction.

This paper takes up some key problems of constructing stochastic models for hydrological processes observed in a number of points in a geographical region. To determine the required statistical characteristics, it is reasonable to average not only over time but also over space (i.e. over observation points). Thus, the corresponding models are referred to as regional models.

Hydrological observations are usually correlated in space and time. As a rule (but not always), a sample consisting of N dependent observations has less information and the estimation accuracy is lower than in the case of independent observations. For example, the information contained in the time series consisting of about 100 mean annual levels of the Caspian Sea is equivalent to that of less than 10 independent observations (see, for example, [2]). It is obvious that statistical estimations based on such data are extremely unreliable. It is important to find out what the number of independent observations N_e, equivalent to the sample of N dependent values, really is.

The necessity of using N_e arises when dealing with many problems and, in particular, when evaluating the mean square error or the number of degrees of freedom, for example, to establish confidence limits or test statistical hypotheses. The data set may consist of either an individual realization of a random process (field) or of an ensemble of observation series for a number of neighbouring points. A detailed discussion of all problems related to this subject can be found in Yaglom [3].

M.V. Bolgov et al. (eds.), Hydrological Models for Environmental Management, 67–73.
© *2002 Kluwer Academic Publishers. Printed in the Netherlands.*

In the following it is assumed that the time series are statistically stationary and have a stationary correlation which taken together form a multidimensional stationary random series. The model discussed contains a description of the statistical structure of K observed time series by means of the appropriate mean value m and covariance matrix $R_{ij}(k_1, k_2)$; $k_1, k_2 = 1, 2, ..., K$, where the value m is constant and $R_{ij} = R_{i-j}$ is a function of only the difference in time i and j.

Formulas for calculation of N_e corresponding to the commonly used estimators m^* and R_{ij}^* are presented in Section 2, where some related topics are also considered.

Numerical values for N_e are presented in Section 3, assuming that the covariance matrix has the simplest and commonly used form. The difficulties arising when the true statistical characteristics are unknown are discussed Section 4. In Section 5 the procedures of regional averaging and constructing regional models are considered. An example demonstrating a noticeable increase in N_e as a result of pooling of the mean annual runoff observations for two stations is given in Section 6. Finally, Section 7 introduces the concept of an equivalent stationary and spatially homogeneous regional model and discusses the problem of its reliability.

2. Formulas for the equivalent number of independent observations N_e

Here the definition of an equivalent (or effective) number of independent observations, N_e, is used which is useful in the majority of cases. To determine this number N_e, only a minimum of information about the statistical structure is needed with no comprehensive knowledge about the probability distribution being required. For a detailed discussion and references see, for example, [1].

If, for example, an estimator m^* of an unknown parameter m is the arithmetic mean of the observation values x_i, i= 1, ..., N, the value of N_e (rather than N) shows how much lower the variance of this estimate is (and the estimation accuracy is thus higher) than that for a single value x_i. This characteristic feature of the quantity N_e can be taken as its definition. Given the value of N_e, the well known classical formulas used in the case of independent observations can be applied for the dependent ones by substituting N_e for N. For example, to determine the number of degrees of freedom or a confidence interval, N_e must be used instead of the sample size N.

The quantity N_e depends not only upon the sample considered but also on the parameter to be estimated. For the estimate m^* we have:

$$N_e(m^*) = (N \, tr \, (\mathfrak{R})) \, / \, (\sum_{i=1}^{N} \sum_{j=1}^{N} R_{ij}) \qquad (1)$$

where $\mathfrak{R} = \{R_{ij}\}$ is the covariance matrix of x_i, i= 1, ..., N.

For the estimates R_{ij}^*, the formulas for $N_e(R_{ij}^*)$, i, j = 1, ..., N differ from eq. (1). Assuming that x_i follow a Gaussian distribution, the formulas for $N_e(R_{ij}^*)$, i, j = 1, ..., N can be easily obtained. In particular, for the estimate of the variance R_0^* we have:

$$N_e(R_0^*) = N^2 R_0^2 \, / \, (\sum_{l=1}^{N} \sum_{j=1}^{N} R_{ij}^2) = (tr(\mathfrak{R}))^2 / \, tr(\mathfrak{R}^2) . \qquad (2)$$

Equation (2) is valid for the case when the mean value m is known. If N is not too small, it can be used also when the mean value is estimated from observation data.

It should be noted that if the sum of all the off-diagonal elements of the matrix in the denominator on the right side of eq. (1) is negative, then the value of N_e exceeds that of N. It is easy to see from eq. (2) that this is not the case for the estimate R_0^* if x_i are Gaussian since all the summands in the denominator are non-negative.

3. An example of the simplest covariance matrix $\mathfrak{R}$

If the covariance matrix is of the form $R_{ij} = R_{i-j} = R_0 \, r^{|i-j|}$, 0< r <1, then the value of N_e (m*) in eq. (1) is considerably lower than the value of $N_e (R_0^*)$ in eq. (2). As the value of r increases, the ratio of the latter one to the former increases as well AS approaching 2. The discrepancy between N_e (m*) and $N_e (R_0^*)$ is most significant when r = 0.5 , as can be seen in Table 1. It is of interest to compare these two values with the value of N_e (R_1^*) (see Table 1).

If the variables x_i are non-Gaussian (as is often the case in hydrology), the number N_e may differ noticeably from the one obtained using such formulas as eq. (2).

This was shown in 1970 by Alekseev and discussed in [3]. Nevertheless, we may expect that these formulas give us the order of magnitude that might be rather helpful in different applications.

TABLE 1. Values of the ratio $N_e(R_0^*)/N$ for different estimators and some values of r

r	.1	.2	.3	.4	.5	.6	.7	.8	.9
$N_e(m^*)/N$	.82	.67	.54	.43	.33	.25	.18	.11	.05
$N_e(R_0^*)/N$	.98	.92	.83	.72	.60	.47	.34	.22	.10
$N_e(R_1^*)/N$	.96	.86	.73	.60	.48	.38	.28	.19	.10

4. A real situation when true statistical characteristics are unknown

Formulas like eq. (1) and eq. (2) can be used if the true statistical characteristics of m and R_{ij} are known. However, it is these particular characteristics that will be estimated here. Thus, in this case the above mentioned formulas strictly speaking are only theoretical. However, not having any additional information, the only solution that remains is to use the sample values of m* and R_{ij}^* instead of the true ones.

It is a common situation in practice that only a sample containing several short observation series is available, and all the statistical parameters, including those needed for estimating mean square errors or confidence intervals, are to be estimated. This gives rise to new difficult problems, which can be considered as key topics in applied statistics.

Referring to the consideration above, of particular interest are the characteristics like $S(m) = \sum_{i=1}^{N} \sum_{j=1}^{N} R_{ij}$ or $S(R_0) = \sum_{i=1}^{N} \sum_{j=1}^{N} R_{ij}^2$ (the denominators in eqs. (1) and (2)) and their sample estimates, $S^*(m)$ and $S^*(R_0)$. It is these cumulative quantities (rather than the individual matrix elements R_{ij} or R_{ij}^2), along with the matrix traces in the numerator, that offer a sufficient base for the computation of N_e. At the same time, characteristics like S^* can be described in terms of x_i by much simpler formulas than those for N_e, and therefore, they are more suitable for analyses. It is worth noting that characteristics like S^* are suitable as indicators of the positive definiteness of the covariance matrices of a sample, and can be helpful when determining a truncation point I_0 for the temporal covariance function $\overset{*}{R}_i$ of a sample (after which zeros are substituted for the truncated tail: $\overset{*}{R}_i = 0, i > I_0$).

5. Regional averaging

It is quite common in hydrological applications to have an ensemble of time series x_i (k), k=1,2,..., K, i=1, 2,..., N_k obtained, say, at K neighbouring observation points. As a rule, estimates calculated from any individual time series are not reliable because the series available are rather short and determining individual statistical parameters for a separate time series makes no sense. However, if the physiographical conditions are more or less similar for the observation points considered, it is reasonable to assume that the statistical properties of the corresponding time series are also approximately the same. This fact allows pooling of the K time series x_i (k) to one sample, which results in a noticeable increase in the amount of information for estimation of their (common) mean value m and auto-covariance function R_{i-j} (k, k)= R_{i-j}. The degree of this increase, however, is considerably smaller than K due to significant positive cross-correlations between the observations in the neighbouring points. The value of N_e calculated for the pooled ensemble of the K time series can specify numerically the degree of this increase. Under the assumption of stationarity, we must assume that not only the auto- but also all the cross-covariance functions depend only on the difference in time, i.e. R_{ij} (k_1 , k_2)= R_{i-j} (k_1 , k_2).

To estimate the values m and R_{i-j} , it is convenient to arrange all the time series $x_i(k)$, k=1,2,..., K sequentially and number them from 1 to   N= $N_1 + N_2 + ...+ N_K$. As a result, we obtain a series of N observations having the constant mean value m and the covariance matrix consisting of the matrix blocks, where the auto-covariance matrices R_{i-j} (k, k)= R_{i-j} , i, j = 1,..., N_k are along the main diagonal and the cross-covariance matrices (generally speaking not square but rectangular) are represented by the off-diagonal blocks. From the above discussion it follows that we may use the empirical (i.e. estimated) values $\overset{*}{R}_{i-j}(k_1, k_2)$ in order to calculate N_e for the estimates m* and $\overset{*}{R}_{i-j}$ with the help of eqs. (1) and (2). (Note that the basic composed covariance matrix should be appropriately reduced at the common points of the adjacent time series depending on the time lag value i). In hydrological applications all the spatial cross-covariances, except for the synchronous ones (i=j) and perhaps those for the adjacent time moments (i-j= 1 or i-j=-1) can be regarded as zeros.

Up to this point the covariance functions have been considered. In hydrology, however, correlation functions (i.e. normalized by the variance) $r_{i-j} = R_{i-j}/R_o$ are used more often. The matter is that, even when the physiographic conditions are similar, the mean magnitudes of the observed values may differ significantly for different observation points due to different catchment areas. It is possible to assume that the statistical parameters are approximately the same for different points only after normalizing by the corresponding standard deviation. It is obvious that for such normalized values the covariance and correlation functions coincide and $r_o = R_0 = 1$.

6. An example for the case with two observation points

Consider the simplest example of two time series of river runoff observed at two neighbouring points. Assume that the true correlation function is the same for both time series and is equal to the arithmetic mean of their empirical correlation functions. Table 2 shows the equivalent number of observations N_e for the autocorrelation coefficients for the case when they are estimated separately from each of the time series ($r_1(1, 1)$ and $r_1(2, 2)$) and when the estimate is obtained from the pooled sample (r_1). It can be seen that pooling leads to a significant increase in the amount of information N_e and consequently to the higher accuracy of estimation. The equivalent number of observations N_e for the cross-correlation coefficients are presented also at the 4-th, 5-th, and 6-th rows of Table 2.

Table 2. Values of N_e for different estimated parameters when using each of the time series separately and the pooled sample

Estimated parameter	Empirical value r_i^*	Time series length N	Equivalent number N_e
$r_1(1, 1)$	.36	86	69
$r_1(2, 2)$	.42	41	34
r_1	.39	127	104
$r_0(1, 2)$	.55	42	32
$r_1(1, 2)$	.42	41	33
$r_1(2, 1)$	.11	41	32

7. Concept of an equivalent stationary and spatially-homogeneous stochastic model

Trying to increase the sample size, some additional assumptions about the statistical structure of the available time series have to be made. Here, it is assumed that the mean values and autocorrelation functions are statistically stationary and spatially homogeneous. Naturally, we can neither prove nor reject the statistical stationarity and spatial homogeneity of the ensemble of observations. Thus, the regional stochastic model obtained as a result of pooling described above can be considered as an *equivalent stationary and spatially homogeneous* stochastic model. Such a model satisfactorily describes the statistical parameters averaged over time and general for the whole region. We may hope that these parameters are close enough to their true values for every individual time series. Anyway, the data available, as a rule, do not give a sufficient reason for rejecting the hypothesis about the adequacy of such a model. It seems that creation of such models is the only real way to describe reasonably the ensembles of rather short time series observed.

8. References

1. Fortus, M. I. (1999) Equivalent number of independent observations: a review, *Izv.Acad. Sci., RAN, Atmos. and Oceanic. Phys.*, **35(6)**, 655-662 (in Russian).
2. Prival'sky, V. E. (1985) *Variability of climate (stochastic models, predictability, spectra)*, Nauka, Moscow (in Russian).
3. Yaglom, A. M. (1987) *Correlation theory of stationary and related random processes*, Vol. 1 & 2, Springer-Verlag, New York - Berlin.

A METHOD FOR SIMULATION OF PERIODIC HYDROLOGICAL TIME SERIES USING WAVELET TRANSFORM

GUOZHANG FENG
College of Water Resources and Architectural Engineering
Northwestern Agricultural University
Yangling, Shaanxi, 712100, CHINA

Abstract

Wavelet transform is a new method of mathematical analysis. It can be used to decompose a signal into independent contributions of both time and frequency without losing information in the signal and is regarded as a mathematical microscope. In this paper, a method to simulate periodic hydrological time series (phts) is developed, which is called main periodic random combinatorial reconstruction (MPRCR) method. The MPRCR method is applied to generate monthly, ten-day, five-day and daily streamflow series for five hydrological gage stations in the Wei River basin, the largest tributary of the Yellow River. Comparing to measured series, the simulated series preserves main statistic features. Decomposition and reconstruction of the streamflow series are completed before the generation to analyze the applicability of wavelet transform to phts. It is shown through this study that the MPRCR method has the property of non-parametrization and may be capable of generating either stationary or non-stationary phts. The method may be applicable to simulate other periodic time series.

1. Introduction

Long-term *periodic hydrological time series* (phts) with different time intervals is commonly needed in water resources planning and management. For example, long-term daily or monthly streamflow data are usually required in reservoir operation studies. However, it is often difficult to meet the needs because the available hydrological records are relatively short. Therefore, artificial synthesis of the phts becomes a hopeful way.

75

M.V. Bolgov et al. (eds.), Hydrological Models for Environmental Management, 75–90.

The most popular method for artificial synthesis or simulation of long-term phts is the technique of stochastic or random simulation, i.e. the technique of time series analysis. Typically, the seasonal autoregressive moving average models [1] and disaggregation models [2] have been well developed and widely used in the simulation of long-term phts [3, 4, 5]. These kinds of models have their own merits and are still dominant in the simulation. However, they certainly have some limitations and are usually insignificant effective and even difficult in the simulation of long-term phts with short time intervals such as long-term daily hydrological processes. There is continuous encouragement for developing some more effective methods for the simulation of the phts.

It may be possible to generate phts using the technique of wavelet transform [6]. This idea appears under the enlightenment of the theory of decomposition and reconstruction in wavelet analysis or multi-resolution analysis. It is recognized from the technique of wavelet transform of multi-resolution analysis that measured hydrological time series (one-dimensional signals) can be decomposed into several different signal components (series) with different frequencies or scales and then the decomposed series may be reconstructed with an appropriate manner. The so called appropriate manner means that proposed methods for reconstructing the hydrological time series should be capable of generating longer phts and keeping either the inherent statistic features of the measured series or stochastisity. As a primary study Feng [6] used wavelet transform to simulate hydrological time series. The results were encouraging, especially those of phts simulation. The purpose of this paper is to develop a method for long-term phts simulation using the technique of wavelet transform.

2. Wavelet Transform

Wavelet transform is an important part of the new branch of mathematics, wavelet analysis or multi-resolution analysis. It is able to provide time and frequency information simultaneously in a time series and give a time-frequency representation of the series [7, 8]. In wavelet transform wavelets play an important role. A wavelet is a function that satisfies certain mathematical requirements and can represent time series or other functions. All wavelets are based on a function called mother wavelet or analyzing wavelet $\psi(t)$, which has some special properties and satisfies the condition $\hat{\psi}(0) = 0$. Under the condition of $\hat{\psi}(0) = 0$, wavelet transform for one-dimensional signal (time series) $x(t) \in L^2(R)$ is defined as follows [9].

$$W(\tau,s) = \int_{-\infty}^{+\infty} x(t)\frac{1}{\sqrt{|s|}}\psi^*\left(\frac{t-\tau}{s}\right)dt, \qquad \tau \in R \text{ and } s \in R-0 \qquad (1)$$

where $W(\tau,s)$ is the wavelet coefficient i.e. decomposed signal series; $\psi^*(t)$ is complex conjugate of $\psi(t)$; τ is the position parameter and s is the scale parameter.

The inverse transform of Equation 1 or reconstruction of $x(t)$ can be represented as

$$x(t) = \frac{1}{C_\psi} \int_{-\infty}^{+\infty}\int_{-\infty}^{+\infty} W(\tau,s)\frac{1}{s^2}\psi\left(\frac{t-\tau}{s}\right)d\tau\,ds \qquad (2)$$

where C_ψ is the constant of admissibility, which depends on the wavelet used and satisfies the condition of admissibility

$$C_\psi = \int_{-\infty}^{+\infty}\frac{|\hat{\psi}(\omega)|^2}{|\omega|}d\omega < \infty \qquad (3)$$

where $\hat{\psi}(\omega)$ is the Fourier transform of $\psi(t)$.

In practice, it is necessary to discrete wavelets and their transformations to suit for certain algorithms. By binary discretization, Equation 2 can be written as

$$x(t) = \sum_{j\in Z}\sum_{k\in Z} \varphi_{jk}\psi_{jk}(t) \qquad (4)$$

where φ_{jk} is the wavelet coefficient, $\varphi_{jk} = W(\tau,s) = W(k, 2^j)$; and

$$\psi_{jk}(t) = \frac{1}{\sqrt{2^j}}\psi\left(\frac{t-2^j k}{2^j}\right) \qquad (5)$$

In this paper, the so called compactly supported quadratic spline function is adopted as the mother wavelet function $\psi_{jk}(t)$, i.e. $\psi(t)$ [10, 11], and the Mallat's algorithm of multi-resolution decomposition and reconstruction of binary wavelets [12] is used.

3. Concept of Simulation of phts by Wavelet Transform

In the application of wavelet analysis such as signal processing and pattern recognition, a signal series is decomposed into several series of wavelets with

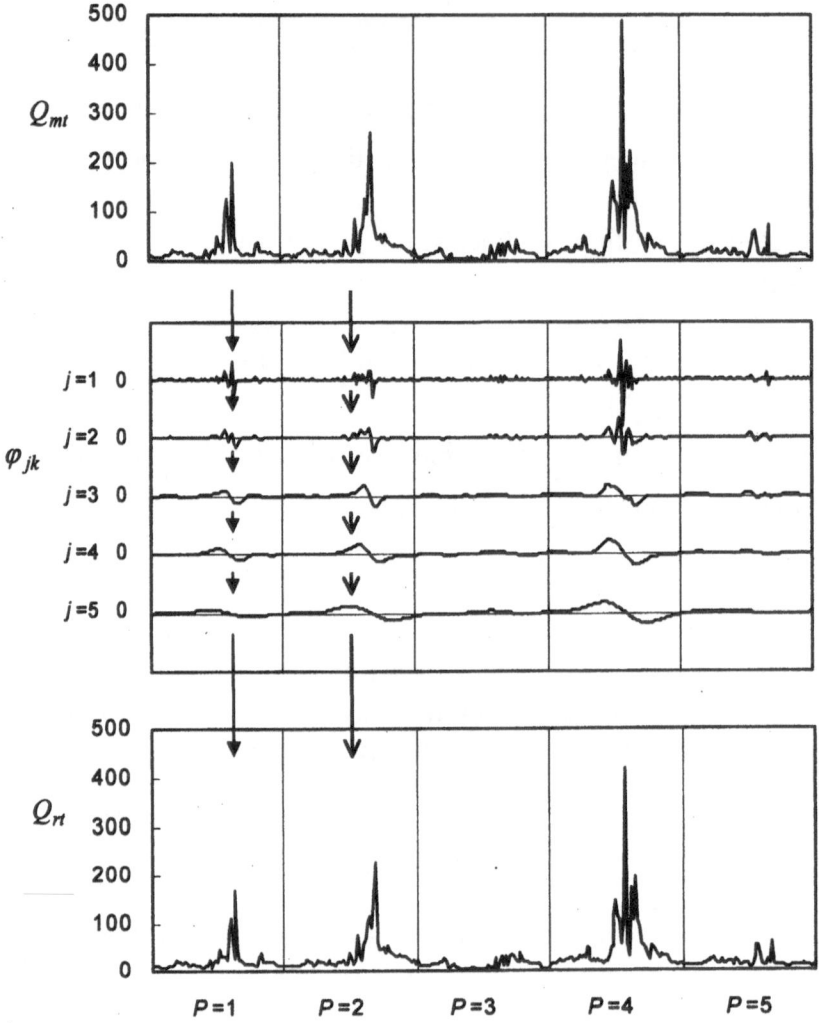

Figure 1. Illustration of traditional wavelet decomposition and reconstruction: (a) a measured time series with five main periods (here it is a ten-day average discharge series Q_{mt}), (b) decomposed wavelet series φ_{jk} and (c) the reconstructed series Q_{rt}, in which the processes from Q_{mt} to x_t and from y_t to Q_{rt} are omitted.

different resolutions or decomposition scales and then reconstructed using the technique of wavelet transform. The decomposition and reconstruction are in one-to-one correspondence (Figure 1). Therefore, the reconstructed series is a restoration of the original series [10, 11].

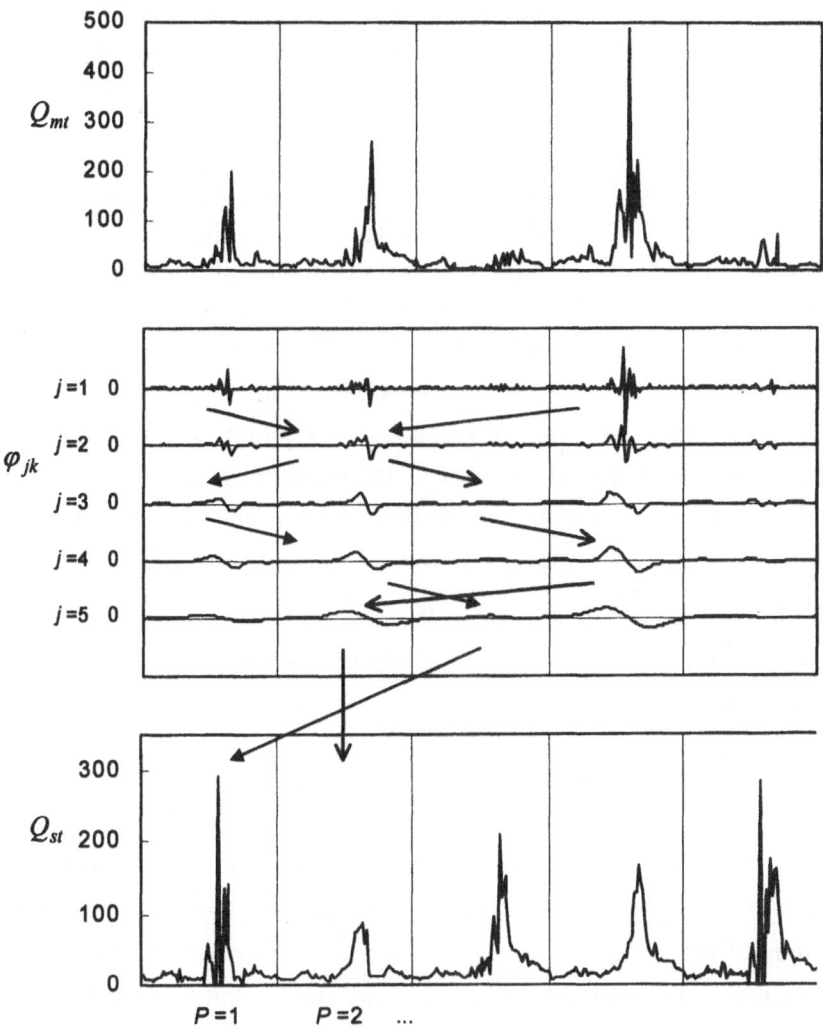

Figure 2. Illustration of the concept of the main period random combinatorial reconstruction of phts: (a) measured time series with five main periods (here it is a ten-day average discharge series Q_{mt}), (b) decomposed wavelet series φ_{jk} and (c) simulated series Q_{st}, in which the processes from Q_{mt} to x_t and from y_t to Q_{st} are omitted. The arrays show some samples of the random selection in the combinatorial reconstruction.

It is obvious that as discrete signals on time scale phts can be decomposed and reconstructed by wavelet transform. The problem is how to reconstruct the decomposed signals. A possible approach to solve the problem is: (1) to decompose a phts into wavelet series with different resolutions; (2) to divide the decomposed wavelets into several short series one each period; and (3) to reconstruct the short series with an appropriate manner to generate a longer phts.

In this paper, a method of random combination of the short wavelet series according to their resolutions is used for the reconstruction. Figure 2 illustrates this basic concept, in which the simulated series for a period may be reconstructed by the decomposed signals at different scales through an arbitrary combination of the periods. The concept may be further described as follows.

Let $x_t(t=1,2,\cdots,n)$ be a phts with P significant periods and length T each period, i.e. $T_j(j=1,2,\cdots,P)$. A significant period in the phts is assumed to be a year and is called main period hereafter. The length T of the main period depends on time intervals of the phts or the number of seasons in a year. For instance, $T=12$ in a monthly streamflow series, and $T=36$ in a ten-day streamflow series. Furthermore, let $w_{tj}(t=1,2,\cdots,n; j=1,2,\cdots,S)$ stand for the wavelet series decomposed from the series x_t with resolution 2^j under the decomposition scale $j=1,2,\cdots,S$, then a longer phts $y_t(t=1,2,\cdots,N)$ may be generated as follows

$$y_t = f_{T_m}(w_{t1}^{T_i},\cdots,w_{t2}^{T_k},\cdots,w_{tS}^{T_l}) \qquad (6)$$

where $T_m(m=1,2,\cdots,M)$ stands for the m-th period in the simulated phts $y_t(t=1,2,\cdots,N)$ in which N is the length of the phts and M is the number of the periods of the phts; $T_i,\cdots,T_k,\cdots,T_l$ $(i,\cdots,k,\cdots,l \in P)$ stand for the order of the period in the wavelet series w_{tj} and are randomly selected from the decomposed series but their subscripts $i,\cdots,k,\cdots,l$ do not equal each other simultaneously in order to avoid the appearance of the original records in the simulated phts.

Accordingly, if a decomposition scale S is given for the series x_t, then the length of the simulated phts is

$$N = TP^S - n = n(P^{S-1}-1) \qquad (7)$$

and the number of periods in the simulated phts is

$$M = \frac{N}{T} = \frac{n}{T}(P^{S-1}-1) \qquad (8)$$

It means that for a monthly streamflow series with a length of 50 years, i.e. $P = 50$, $T = 12$, and $n = PT = 600$, a long-tern monthly streamflow series with $N = 600 \times (50^2 - 1) = 1,499,400$ months or $M = 124,950$ years can be generated, assuming $S = 3$.

This method is therefore named as main periodic random combinatorial reconstruction (MPRCR) method according to its major properties.

4. Decomposition and Reconstruction of Hydrological Time Series

As an application of the proposed method, monthly, ten-day, five-day and daily streamflow records with certain one-year main period at five hydrological gage stations in the Wei River basin, the largest tributary of the Yellow River, are used to generate long-term phts for the stations. The relevant information about the five selected hydrological gage stations is shown in Table 1.

In this paper, the ten-day and five-day stand for two Chinese words, *xun* and *hou*, respectively. The *xun* means that the number of days in the first two ten-days in a month is exactly 10 days, and in the last ten-day the number of days is 10 days in April, June, September and November, 8 or 9 days in February, and 11 days in other months. Similarly, the *hou* means that the number of days in the first five five-days in a month is exactly 5 days but varies among 3 to 6 days depending on the month and year under consideration.

The lengths of the original series for the four time intervals used for this study are 432 months (1951-1986), 540 ten-days (1951-1965), 576 five-days (1951-1958) and 1095 days (1981-1983), respectively. The difference in the period P for different time intervals is partly due to the limitation of the computer capacity used in this study.

For the purpose of keeping the phts in a similar magnitude the data are standardized prior to the decomposition as

TABLE 1. Information on the selected hydrological gage stations

Station name	River name	Catchment area (km^2)
Zhuangtou	North Luo River (tributary of the Wei River)	25,154
Zhangjiashan	Jing River (tributary of the Wei River)	43,216
Linjiacun	Wei River	30,661
Xianyang	Wei River	46,827
Huaxian	Wei River	106,498

NB: total catchment area of the Wei River basin is 134,766 km^2.

$$x_t = \frac{Q_{mt} - Q_o}{\sigma} \tag{9}$$

where Q_{mt} is measured series of mean discharges; Q_o is the mean of Q_{mt}; and σ is standard variance of Q_{mt}. All the discharges and related values are in $m^3 s^{-1}$ in this paper.

Similarly, the generated series y_t should be transferred into the mean discharges Q_{rt} as

$$Q_{rt} = Q_o + \sigma y_t \tag{10}$$

The standardized discharge series of the four time intervals are decomposed with $S = 1,2,\cdots,5$ and then reconstructed as usual for comparison to the measured series. Furthermore, the basic statistics of the reconstructed series are compared to those of the original records. The compared statistics are Q_o, σ, C_v (coefficient of variance), C_s (coefficient of skewness) and ρ_1 (lag-one correlation coefficient). The results indicate that the reconstructed series basically preserve the statistics well with completely equal values of Q_o, a slight decrease in σ or C_v and C_s, and increase in ρ_1 with the increase of S. Table 2 shows the difference in the statistics at two of the five stations, Zhuangtou and Huaxian.

Figure 3 illustrates a comparison of reconstructed to measured data for two time intervals, month and ten-day, at the Zhuangtou gage station. The results show that the reconstructed discharges Q_r are less than the measurements Q_m in the range of high values and inverse in the range of low values, in which Q_r and Q_m stand for Q_{rt} and Q_{mt} while without considering the time. The difference trends to be significant with the increase of S.

5. Simulation of Long-term Periodic Hydrological Time Series

It is obvious that for long-term phts simulation the larger the scale S, the longer the simulated phts to a given length of the decomposed series. However, as mentioned above, the larger the scale S, the larger the difference between the simulations and the measurements. Meanwhile, it is impossible for $S = 1$ to generate a phts which is longer than the decomposed series. Therefore, $S = 2$ is used in the simulation of long-term discharge series under the adopted mother wavelet function in this study using the MPRCR method. In the cases, $P = 36$ and $T = 12$ for monthly series, $P = 15$ and $T = 36$ for ten-day series, $P = 8$

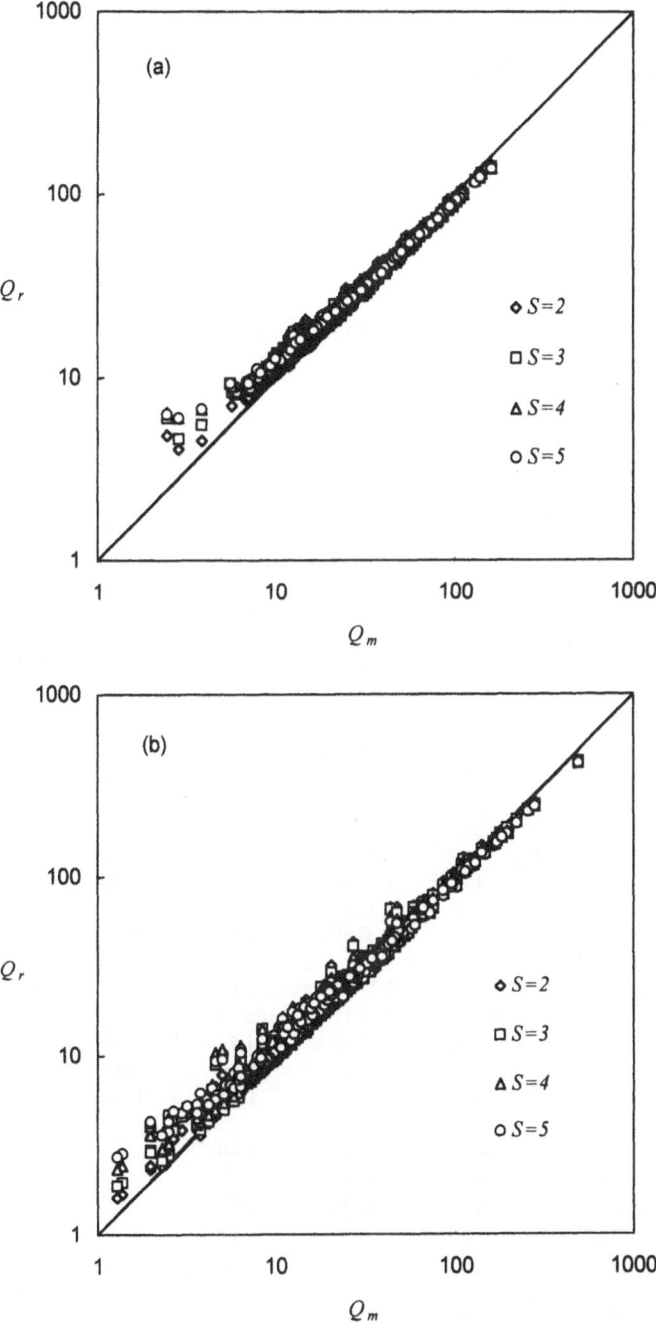

Figure 3. Comparison of the reconstructed time interval mean discharges Q_r with different scale S to the measurements Q_m at Zhuangtou: (a) monthly mean discharges and (b) five-day mean discharges.

TABLE 2. Statistics of the measured and reconstructed streamflow series

Station name	Statistics	Measured series	Reconstructed series				
			$S=1$	$S=2$	$S=3$	$S=4$	$S=5$
Monthly series							
Zhuangtou	Q_o	28.28	28.28	28.28	28.28	28.28	28.28
	σ	23.66	22.52	21.64	20.95	20.66	20.55
	C_v	0.84	0.80	0.77	0.74	0.73	0.73
	C_s	2.50	2.42	2.41	2.44	2.44	2.45
	ρ_1	0.58	0.65	0.64	0.63	0.62	0.61
Huaxian	Q_o	256.17	256.17	256.17	256.17	256.17	256.17
	σ	275.52	260.27	249.87	241.89	239.56	238.23
	C_v	1.08	1.02	0.98	0.94	0.94	0.93
	C_s	2.21	2.10	2.10	2.14	2.14	2.15
	ρ_1	0.52	0.60	0.59	0.57	0.57	0.56
Ten-day series							
Zhuangtou	Q_o	28.01	28.01	28.01	28.01	28.01	28.01
	σ	30.66	29.63	29.05	28.34	27.58	27.01
	C_v	1.10	1.06	1.04	1.01	0.99	0.96
	C_s	3.23	3.09	3.07	3.06	3.09	3.12
	ρ_1	0.70	0.77	0.77	0.76	0.75	0.74
Huaxian	Q_o	289.78	289.78	289.78	289.78	289.78	289.78
	σ	334.83	321.23	312.86	305.11	296.76	292.08
	C_v	1.16	1.11	1.08	1.05	1.02	1.01
	C_s	2.72	2.59	2.56	2.56	2.61	2.65
	ρ_1	0.64	0.71	0.72	0.71	0.69	0.68
Five-day series							
Zhuangtou	Q_o	25.89	25.89	25.89	25.89	25.89	25.89
	σ	37.58	35.67	35.13	34.59	33.81	33.05
	C_v	1.45	1.38	1.36	1.34	1.31	1.28
	C_s	5.78	5.23	5.18	5.22	5.29	5.41
	ρ_1	0.55	0.65	0.66	0.65	0.63	0.62
Huaxian	Q_o	283.63	283.63	283.63	283.63	283.63	283.69
	σ	381.29	365.11	356.71	349.22	341.94	333.72
	C_v	1.34	1.29	1.26	1.23	1.21	1.18
	C_s	3.60	3.35	3.29	3.29	3.32	3.41
	ρ_1	0.62	0.71	0.71	0.70	0.69	0.67
Daily series							
Zhuangtou	Q_o	30.39	30.39	30.39			
	σ	28.92	28.54	28.92			
	C_v	0.95	0.94	0.95			
	C_s	2.89	2.80	2.89			
	ρ_1	0.88	0.90	0.88			
Huaxian	Q_o	298.90	298.90	298.90			
	σ	518.31	512.08	505.46			
	C_v	1.73	1.71	1.69			
	C_s	4.24	4.13	4.06			
	ρ_1	0.89	0.91	0.92			

TABLE 3   Comparison of the statistics of simulated phts to those of the measured phts

Station name	Statistics	Measured series	Simulated series	Error	Relative error (%)
Monthly series					
Zhuangtou	Q_o	28.28	28.34	0.06	0.21
	σ	23.66	19.91	-3.75	-15.85
	C_v	0.84	0.70	-0.14	-16.13
	C_s	2.50	1.73	-0.77	-30.84
	ρ_1	0.58	0.63	0.06	9.50
Huaxian	Q_o	256.17	256.11	-0.06	-0.02
	σ	275.52	225.79	-49.73	-18.05
	C_v	1.08	0.88	-0.19	-18.03
	C_s	2.21	1.48	-0.73	-32.97
	ρ_1	0.52	0.57	0.06	11.07
Ten-day series					
Zhuangtou	Q_o	28.01	28.33	0.32	1.16
	σ	30.66	27.38	-3.28	-10.70
	C_v	1.10	0.97	-0.13	-11.78
	C_s	3.23	2.40	-0.83	-25.69
	ρ_1	0.70	0.79	0.09	13.09
Huaxian	Q_o	289.78	293.73	3.95	1.36
	σ	334.83	286.72	-48.11	-14.37
	C_v	1.16	0.98	-0.18	-15.50
	C_s	2.72	1.86	-0.87	-31.80
	ρ_1	0.64	0.73	0.09	14.24
Five-day series					
Zhuangtou	Q_o	25.89	26.42	0.53	2.05
	σ	37.58	33.09	-4.49	-11.95
	C_v	1.45	1.25	-0.20	-13.71
	C_s	5.78	3.47	-2.31	-40.00
	ρ_1	0.55	0.71	0.16	29.82
Huaxian	Q_o	283.63	288.51	4.88	1.72
	σ	381.29	328.10	-53.19	-13.95
	C_v	1.34	1.14	-0.21	-15.40
	C_s	3.60	2.44	-1.16	-32.32
	ρ_1	0.62	0.74	0.11	17.98
Daily series					
Zhuangtou	Q_o	30.39	30.51	0.12	0.39
	σ	28.92	27.04	-1.88	-6.51
	C_v	0.95	0.89	-0.07	-6.93
	C_s	2.89	2.24	-0.66	-22.77
	ρ_1	0.88	0.93	0.05	5.23
Huaxian	Q_o	298.90	302.40	3.50	1.17
	σ	518.31	485.67	-32.64	-6.30
	C_v	1.73	1.61	-0.13	-7.38
	C_s	4.24	3.46	-0.78	-18.33
	ρ_1	0.89	0.94	0.04	4.97

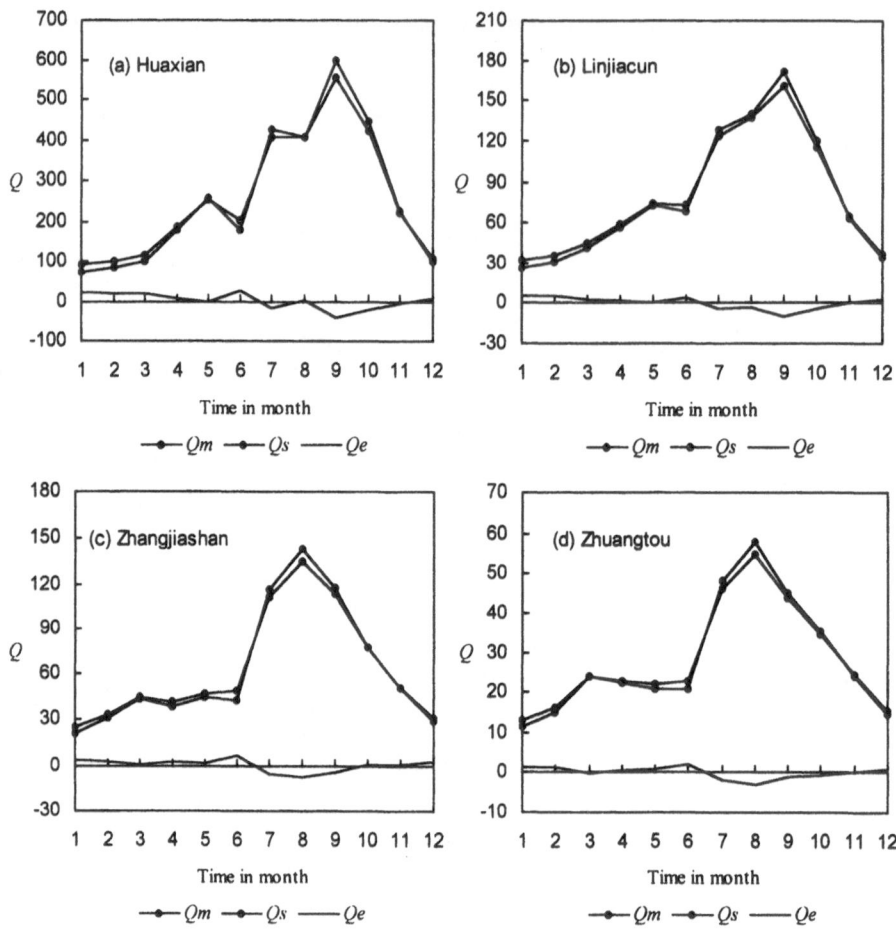

Figure 4. Comparison of in-year distributions of averages of simulated long-term monthly mean discharges (Q_s) to those of the measured discharges (Q_m) at four hydrologic gage stations in the Wei River basin (Q_e stands for the differences between Q_s and Q_m, i.e. $Q_e = Q_s - Q_m$).

and $T = 72$ for five-day series, and $P = 3$ and $T = 365$ for daily series, respectively. It means that the lengths of the generated series are 15,120 months or 1,260 years for the monthly discharges, 7,560 ten-days or 210 years for the ten-day series, 4,032 five-days or 56 years for the five-day series, and 2,190 days or 6 years for the daily series, respectively.

The results show that the simulated series of the discharges for the four time intervals at the five hydrological gage stations can preserve essential statistic properties comparing to the original observations. There is no significant difference between the simulations and observations in the mean

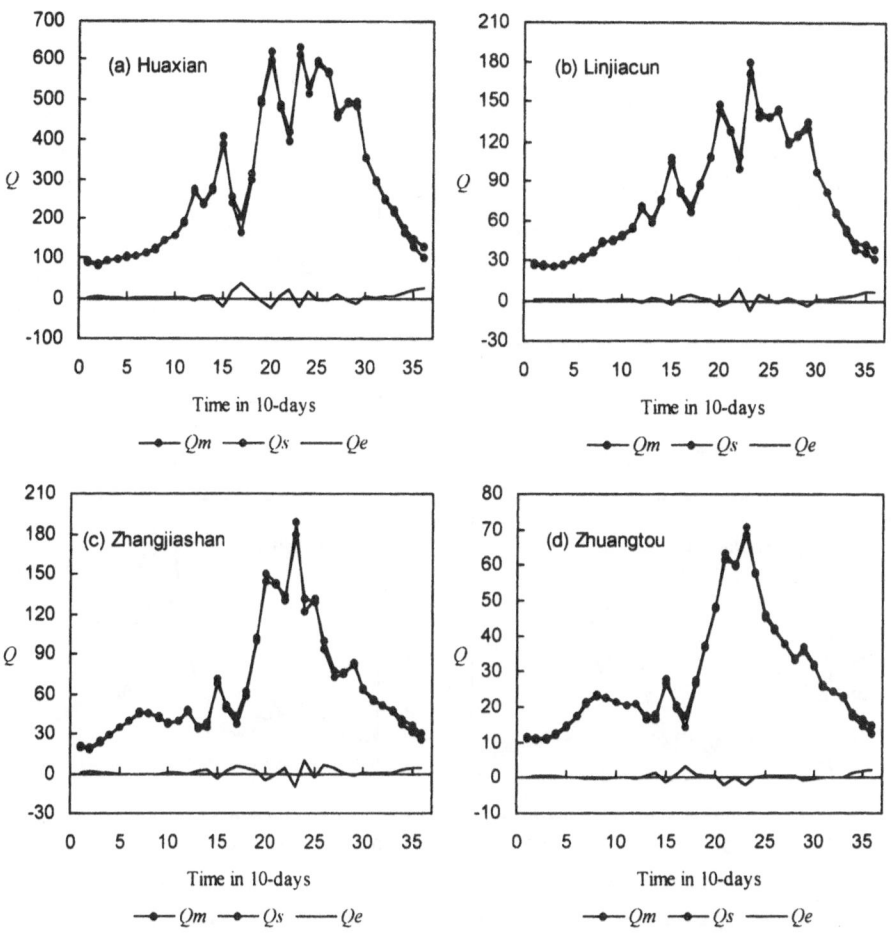

Figure 5. Comparison of in-year distributions of averages of simulated long-term ten-day mean discharges (Q_s) to those of the measured discharges (Q_m) at four hydrologic gage stations in the Wei River basin (Q_e stands for the differences between Q_s and Q_m, i.e. $Q_e = Q_s - Q_m$).

discharges. The maximum relative error in the mean discharges is less than 2%. The standard variances of the simulations are persistently <18% compared to those of the originals for the monthly series, <15% for ten-day and five-day series, and <7% for daily series. The coefficients of skewness are in large errors up to 35% of relative errors. The relative errors in lag-one correlation coefficients are around 15% for most series. Table 3 shows the difference between the statistics of the simulated phts and those of the measured phts, at two of the five stations, Zhuangtou and Huaxian, as examples. Figure 4 to Figure 7 illustrate the difference between the in-year distributions of the mean

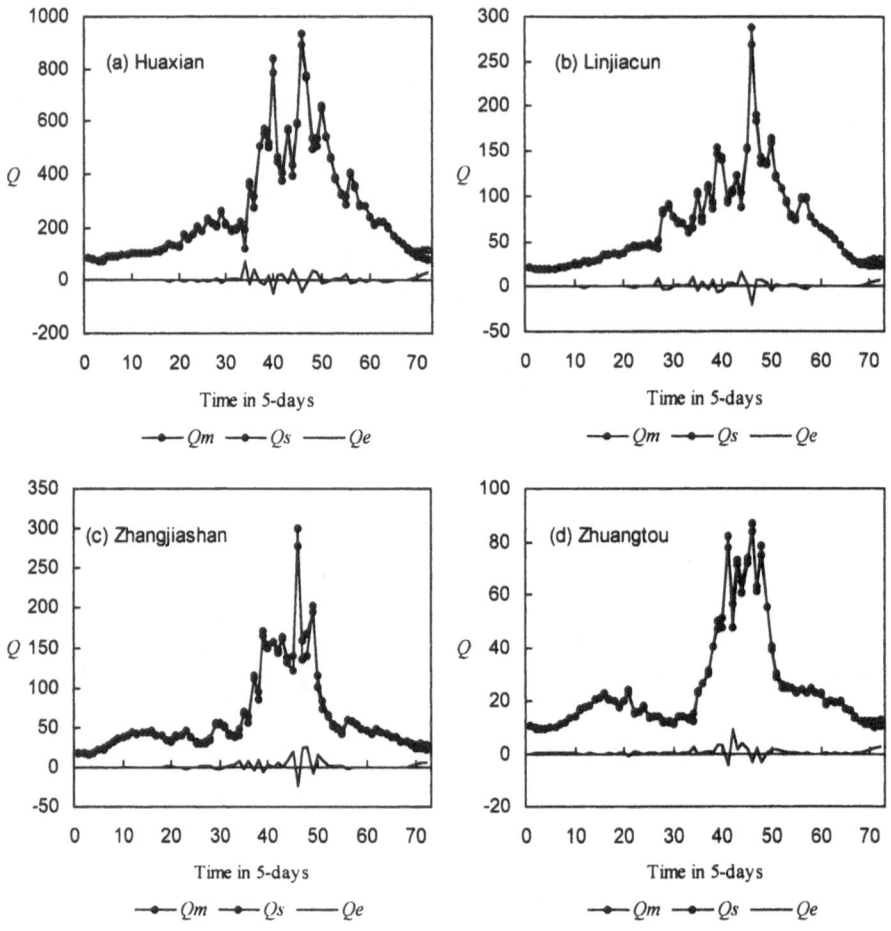

Figure 6. Comparison of in-year distributions of averages of simulated long-term five-day mean discharges (Q_s) to those of the measured discharges (Q_m) at four hydrologic gage stations in the Wei River basin (Q_e stands for the differences between Q_s and Q_m , i.e. $Q_e = Q_s - Q_m$).

discharges of the simulations and the observations for the four time intervals at four of the five stations excepting for Xianyang. It is obvious from these figures that the fitness of the simulated phts to the measured phts on an average level is quite close each other.

6. Discussion and Conclusions

In this paper the main periodic random combinatorial reconstruction (MPRCR)

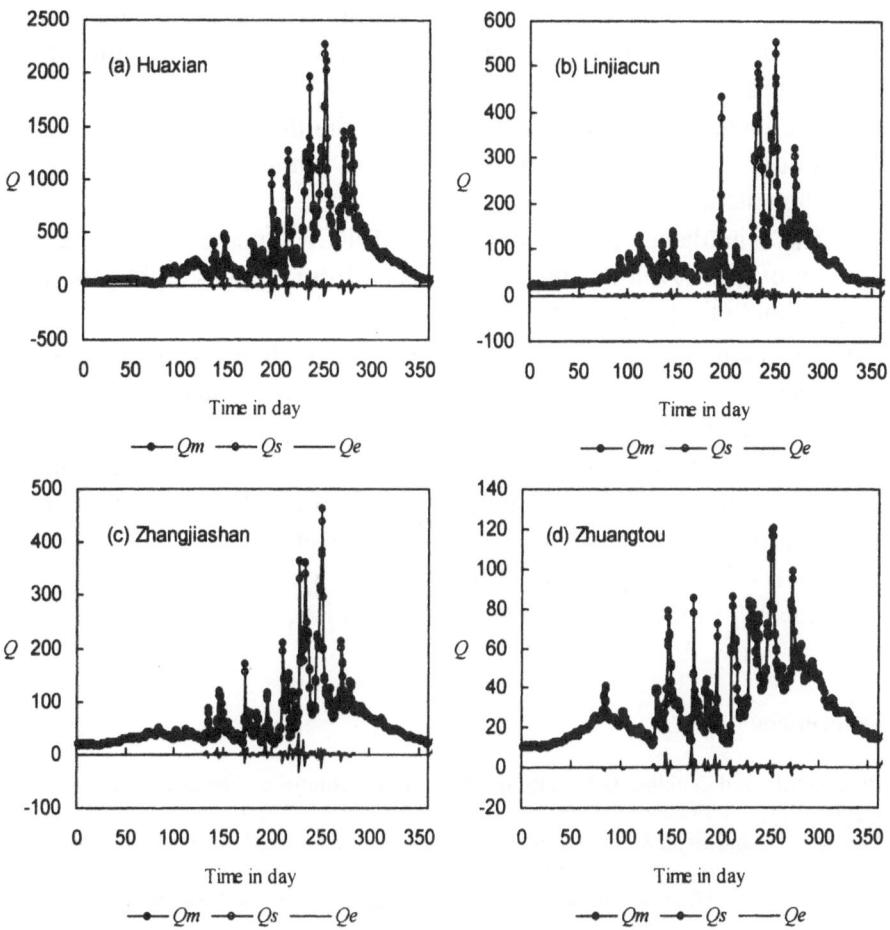

Figure 7. Comparison of in-year distributions of averages of simulated long-term daily mean discharges (Q_s) to those of the measured discharges (Q_m) at four hydrologic gage stations in the Wei River basin (Q_e stands for the differences between Q_s and Q_m, i.e. $Q_e = Q_s - Q_m$).

method was developed for simulation of ling-term periodic hydrological time series (phts). The results from the application of the MPRCR method to generate long-term monthly, ten-day, five-day and daily streamflows shown that the proposed method have several advantages over the seasonal ARMA and disaggregation models, and may be concluded as follows.

(1) The MPRCR method may be suitable for stationary and non-stationary phts without any stationary transformation.

(2) An obvious advantage of the method over the traditional methods of time series analysis is its non-parametrization. The only parameters used for the

simulation are the two basic statistics, mean and variance, for overall standardization of the measured phts and inverse transformation of the simulated standardized phts.

(3) The method is capable of overcoming the drawback that the sum of seasonal values in a year is not equal to the annual value in using the disaggregation models.

(4) The simulated phts using the MPRCR method is a random combination of the wavelet signals decomposed from measured phts, therefore it is a reflection to the reality of the phts.

(5) The method seems to be reasonable and reliable for simulation of time series with significant main period and may be successfully applied to generate long-term phts and other periodic time series.

(6) The persistent lower or higher deviations in the statistics may be caused by the effects of noise elimination of the filter functions used in the transform. Further attention should be focus on development of more effective filter functions or the mother wavelet functions for hydrological time series and other approaches, which could improve the simulation.

7. References

1. Box, G.E.P. and Jenkins, G.M. (1976). *Time Series Analysis: Forecasting and Control*, Holden-Day, San Francisco.
2. Valencia, D. and Schaake, J.C. (1973). Disaggregation processes in stochastic hydrology, *Water Resources Research* **9**, 580-585.
3. Yevjevich, V. (1972). *Stochastic Processes in Hydrology*, Water Resources Publications, Colorado.
4. Bras, R.L. and Rodriguez-Iturbe, I. (1985). *Random Functions and Hydrology*, Addison Wesley Publishing Company, US.
5. Hipel, K.W. and Mcleod, A.I. (1994). *Time Series Modeling of Water Resources and Environmental Systems*, Elsevier Science, The Netherlands.
6. Feng, Guozhang. (1998). A method of wavelet transform for generation of hydrological time series, in Dedun Song (ed.) *The Selected Papers of Symposium on Advances and Prospects for Hydrologic Computation of China*. Hehai University Press, Nanjing (in Chinese).
7. Chui, C.K. (1992). *An Introduction to Wavelet*, Academic Press Inc. Boston.
8. Liu, Guizhong and Shuangliang Di. (1995). *Wavelet Analysis and Its Application*, Press of Xi'an University of Electronic Science and Technology, Xi'an (in Chinese).
9. Grossmann, A. and Morlet, J. (1984). Decomposition of Hardy functions into square integrable wavelets of constant shape, *SIAM J. Math. Anal.* **15**, 723-736.
10. Qin, Qianqing and Zongkai Yang. (1994). *Applied Wavelet Analysis*, Press of Xi'an University of Electronic Science and Technology, Xi'an (in Chinese).
11. Xu, Peixia and Gongliang Sun. (1996). *Wavelet Analysis and Examples of Its Applications*, Press of the University of Science and Technology of China, Hefei (in Chinese).
12. Mallat, S. (1989). A theory for multi-resolution signal decomposition: the wavelet representation, *IEEE Trans on Pattern, Anal and Math, Intel.* **11**, 674-693.

THE CASPIAN SEA AS A STOCHASTIC RESERVOIR

A.V. FROLOV
Water Problems Institute
Russian Academy of Sciences
E-mail: FROLOVS@AHA.RU

1. Introduction

Studies of the variations of the level of natural water bodies are often based on the concept of a "stochastic reservoir". The stochastic reservoir is a physically-based mathematical model of the water level variations in a closed reservoir. In this paper the stochastic reservoir is considered as a hydrological system governed by two input processes *viz.* river inflow and an effective evaporation from the water body (evaporation minus precipitation); a state variable viz. the water level variations; and an output process *viz.* the outflow from a water body. It is assumed that input processes are auto- and mutually correlated. The outflow is a function of the water level variations. This definition of a "stochastic reservoir" is very similar to the one suggested by Lloyd [15].

The theoretical principles of the stochastic reservoir model were developed by Kritsky and Menkel [12], Lloyd [14, 15], Klemeš [10, 11], Kaczmarek [9], Phatarfod [18], Gates and Diesendorf [7], Muzylev [17], Privalsky [19], Ratkovich [20] and others. The stochastic reservoir representation offers a basis for studies of natural hydrological systems like lakes, inland seas, river basins and groundwater aquifers [10,11]. The Caspian Sea was probably one of the first natural water bodies modelled as a stochastic reservoir. Kritsky and Menkel published a paper dealing with the modelling of the water volume variations in the Caspian Sea as early as 1946 [12]. The same approach was also used in many other studies of long-term level variations of the Caspian Sea in [2-4], [8], [13] and [17].

Traditionally the Caspian Sea is considered as a closed lake. However, at least in the 20th century the Caspian Sea had an outflow into the Kara Bogas Gol Bay. The Caspian Sea existed as a closed lake only between 1980 and 1992 when the Bay was cut off by the dam constructed in Kara Bogaz Gol Strait (Fig. 1).

M.V. Bolgov et al. (eds.), Hydrological Models for Environmental Management, 91–107.

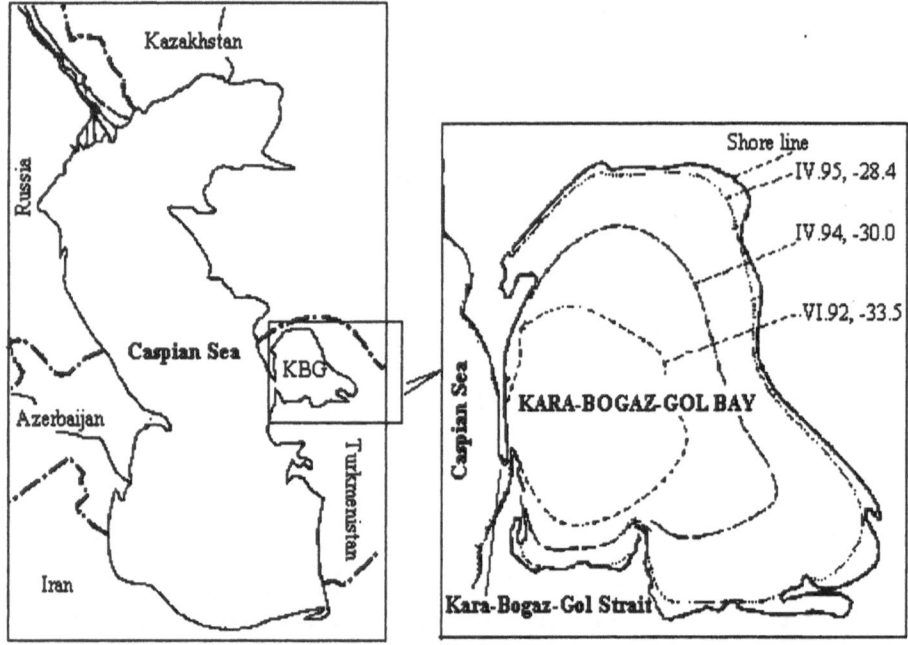

Figure 1. The Caspian Sea (left) and the filling of the Kara Bogaz Gol Bay after June 1992 (right) (after [1]; Roman numerals denote the month, figures to the right are the level marks).

The modelling of the long-term variations in the level of the Caspian Sea has both theoretical and applied aspects. Theoretical research is directed towards understanding the mechanisms behind the variations of Sea level, while the correctly physically-based model allows solution of applied problems such as the estimation of breaks in the Sea level variations caused by anthropogenic and natural impacts. For example, estimation of the probabilities for changes in Sea level in the future was obtained on the basis of the model that will be further discussed below. These estimates were used in the Russian Federal Programme "Caspy" [8] elaborated during the modern catastrophic rise of the Caspian Sea level.

The model has some specific features. First, the river inflow and the effective evaporation from the surface of the Caspian Sea are modeled as serially and mutually correlated stochastic processes, representing the two-dimensional input process. Secondly, the Caspian Sea is considered as an open lake because of the marine water outflow into the Kara Bogaz Gol Bay. Thus, the damping effect of the marine water outflow to the variations in the Sea level is recognized. Thirdly, the model takes into account the non-stationarity in the input processes formed by initial conditions.

2. Basic Equations and Considerations

The variations of water level in an open lake are described by the water balance equation:

$$\frac{dh(t)}{dt} = \frac{v^+(t)}{F(h)} - \frac{v^-(t)}{F(h)} - e(t) \tag{1}$$

where h is the level (m) with respect to the equilibrium level h^*, which is assumed to be zero, $F(h)$ is the area of the lake (in km^2) at level h, $v^+(t)$ and $v^-(t)$ are the river inflow into the lake and the outflow from the lake (both in km^3/year), respectively; and $e(t)$ is the effective evaporation (m/year).

The equilibrium level h^* is the level mark for which the volume of water coming into the lake is equal to the discharge from the lake:

$$< F(h^*)e(h^*) > + < v^- > = < v^+ > \tag{2}$$

where < > denotes statistical averaging.

As long as the up and down level variations, close to the equilibrium level, are considered without any restrictions, the model of the "infinite stochastic reservoir" is applicable.

For many lakes, the dependence between water level and area is assumed to be linear (for the actual task of sea level variations):

$$F = a + bh. \tag{3}$$

If $bh \ll a$, then:

$$F^{-1} \approx \frac{1}{a}(1 - \frac{b}{a}h). \tag{4}$$

In general, the dependence between the outflow from the lake and the water level is not linear. But, if the level variations are not too large, this dependence can be approximated by a linear function:

$$v^-(t) \approx < v^- > + \lambda h(t), \tag{5}$$

where $<v^->$ is the mean value of the outflow, and λ is a coefficient named "the outflow parameter". This coefficient shows the increase (decrease) of the outflow volume when the level increases (decreases) by 1 m.

It follows from Eqs. (1), (4) and (5) that the open lake level variations can be described by the linearized equation:

$$\frac{dh(t)}{dt} + \frac{bv^+(t)}{a^2}h(t) + \frac{\lambda}{a}h(t) - \frac{b<v^->}{a^2}h(t) = \frac{v^+(t)-<v^->}{a} - e(t) \cdot \tag{6}$$

It constitutes a first order stochastic differential equation at the right-hand side (the stochastic input processes $v^+(t)$ and $v^-(t)$) with a stochastic coefficient ($v^+(t)$ at $h(t)$) at the left-hand side. As shown by Muzylev [16], this stochastic coefficient can be substituted by its mean value $<v^+>$ for many natural lakes including the Caspian Sea. Taking into account this consideration, Eq. (6) is written as:

$$\frac{d(h)}{dt} = -\alpha h(t) + \frac{v^+(t)-<v^->}{a} - e(t), \tag{7}$$

where $\alpha = \alpha_1 + \alpha_2$, $\alpha_1 = b<v^+>/a^2$ and $\alpha_2 = \lambda/a - b<v^->/a^2$.

The coefficient α is a quantitative characteristic of the negative feedback into the hydrological system (the open lake) with the input processes (the river inflow and the effective evaporation) and the output processes (the lake level variations and the outflow from the lake). The coefficient α_1 is determined by the dependence between the lake area and the water level. The coefficient α_2 is related to the dependence between the outflow from the lake and the water level. In accordance with Ratkovich [20], the coefficient α is called "the inertial parameter of the lake". Note that the inertial parameter α does not directly depend on the effective evaporation.

Let us introduce the following notations: $v(t)=[v^+(t) - <v^->]/a$, $v(t)=<v>+v'(t)$ and $e(t)=<e>+e'(t)$, where $<v>$ and $<e>$ are the mean values of $v(t)$ and $e(t)$, $v'(t)$ and $e'(t)$ are the deviations from $<v>$ and $<e>$, respectively. Then, as $<v>=<e>$ (the difference between the mean values of the river inflow and the outflow is equal to the effective evaporation), Eq. (6) can be rewritten as:

$$\frac{d(h)}{dt} = -\alpha h(t) + v'(t) - e'(t). \tag{8}$$

Assume that $v'(t)$ and $e'(t)$ form first-order autoregressive processes AR(1):

$$\frac{dv'(t)}{dt} = -r_v v'(t) + w_1(t), \tag{9}$$

$$\frac{de'(t)}{dt} = -r_e e'(t) + w_2(t), \tag{10}$$

where $\gamma_v = -\ln r_v$, $\gamma_e = -\ln \gamma_e$, r_v and r_e are the coefficients of the correlation between $v'(t)$ and $v'(t+1)$, $e'(t)$ and $e'(t+1)$, respectively; $w_{(i)}$ ($i=1,2$) are the mutually correlated non-

Gaussian white noises with the known mathematical expectations $<w_{(i)}>$, the covariance functions $R_{(i)}(\tau)=D_2^{(i)}\delta(\tau)$ and $K_{(i)}=D_3^{(i)}\delta(\tau_1)\delta(\tau_2)$; $\delta(\tau)$ is the delta-function:

$$\delta(\tau) = \begin{cases} \infty, & \tau = 0 \\ 0, & \tau \neq 0 \end{cases},$$

$D_2^{(i)}$ and $D_3^{(i)}$ are the coefficients of the intensity of the white noises, $i=1,2$; and $\tau = \tau_1 = t_1-t_2$, $\tau_2 = t_3-t_2$, t_1, t_2, t_3 are the time moments. Use of the non-Gaussian white noises allows simulation of the river inflow and the effective evaporation as stochastic processes with given asymmetry. Thus, the model processes $v^+(t)$ and $e(t)$ possess the given mean values, variances, coefficients of asymmetry and autocorrelation functions of the exponential type:

$$r(\tau) = e^{-\gamma|\tau|}. \tag{11}$$

It is also assumed that the covariance function $<w_1w_2>$ can be written as:

$$< w_1 w_2 >= D_{ve}\delta(\tau), \tag{12}$$

where $D_{ve}=r_{ve}(\mu_2^v \mu_2^e)^{1/2}(\gamma_v+\gamma_e)$, r_{ve} is the coefficient of the correlation between the river inflow $v^+(t)$ and the effective evaporation $e(t)$, μ_2^v is the variance of the river inflow layer, μ_2^e is the variance of the effective evaporation. In this case the correlation function $r_{ve}(\tau)$ is described as:

$$r_{ve}(\tau) = \begin{cases} r_{ve}e^{-\gamma_v\tau}, & \tau \geq 0 \\ r_{ve}e^{-\gamma_e|\tau|}, & \tau < 0 \end{cases}. \tag{13}$$

It is easy to show that the coefficients r_{ve}, γ_v and γ_e satisfy the inequality:

$$r_{ve} \leq \frac{2(\gamma_v\gamma_e)^{1/2}}{\gamma_v+\gamma_e}. \tag{14}$$

A solution of the system of eqs. (8-10) gives a complete description of the states of the Sea. The conditional expectation of the level $< h(t) >$ (i.e. the mathematical expectation of the level as a function of both the time and the initial values of the level, river inflow and evaporation) is:

$$< h(t) >= C_1 e^{-\alpha t} + C_2 e^{-\gamma_v t} + C_3 e^{-\gamma_e t}, \tag{15}$$

where C_1, C_2, and C_3 are the constants obtained from the initial conditions:

$$C_1 = h_0 + \frac{v_0'}{\alpha-\gamma_v} - \frac{e_0'}{\alpha-\gamma_e}, \quad C_2 = \frac{v_0'}{\alpha-\gamma_v}, \quad C_3 = -\frac{e_0'}{\alpha-\gamma_e}. \tag{16}$$

Here h_0, v'_0, e'_0 are the initial values of the sea level, river inflow and evaporation in terms of their deviations from the mean values of the corresponding processes; r_v and r_e are the coefficients of the correlation between the values $v(t)$ and $v(t+1)$, $e(t)$ and $e(t+1)$, respectively; and α is the inertial parameter of the Caspian Sea.

The conditional variance of the level $\mu_2^h(t)$ is determined by the formula:

$$\mu_2^h = \frac{\gamma_e \mu_2^e}{(\alpha - \gamma_e)^2} A - r_{ve} \frac{(\mu_2^e \mu_2^v)^{1/2}(\gamma_v + \gamma_e)}{(\alpha - \gamma_v)(\alpha - \gamma_e)} B + \frac{\gamma_v \mu_2^v}{(\alpha - \gamma_v)^2} \Gamma, \qquad (17)$$

where r_{ve} is the coefficient of the cross-correlation between the river inflow and evaporation, and:

$$A = 2 \left[\frac{1 - e^{-2\gamma_e t}}{2\gamma_e} - \frac{2\left[1 - e^{-(\alpha + \lambda_v)t}\right]}{\alpha + \gamma_v} + \frac{1 - e^{-\alpha t}}{2\alpha} \right], \qquad (18)$$

$$B = \left[\frac{1 - e^{-(\gamma_v + \gamma_e)t}}{\gamma_v + \gamma_e} - \frac{1 - e^{-(\alpha + \gamma_e)t}}{\alpha + \gamma_e} - \frac{1 - e^{-(\alpha + \gamma_v)t}}{\alpha + \gamma_v} + \frac{1 - e^{-2\alpha t}}{2\alpha} \right], \qquad (19)$$

$$\Gamma = 2 \left[\frac{1 - e^{-2\gamma_v t}}{2\gamma_v} - \frac{2\left[1 - e^{-(\alpha + \lambda_e)t}\right]}{\alpha + \gamma_e} + \frac{1 - e^{-\alpha t}}{2\alpha} \right]. \qquad (20)$$

For the stationary mode (i.e. as $t \to +\infty$), the variance μ_2^h of the level variance is determined by the formula:

$$\mu_2^h = \frac{\mu_2^v}{\alpha(\alpha + \gamma_v)} - r_{ve} \frac{(\mu_2^v \mu_2^e)^{1/2}(2\alpha + \gamma_v + \gamma_e)}{\alpha(\alpha + \gamma_v)(\alpha + \gamma_e)} + \frac{\mu_2^e}{\alpha(\alpha + \gamma_e)}. \qquad (21)$$

The autocorrelation function $r^h(\tau)$ of the level in the stationary mode is described as:

$$r^h(\tau) = \frac{C_1' e^{-\gamma_v |\tau|} + C_2' e^{-\gamma_e |\tau|} + C_3' e^{-\alpha |\tau|}}{C_1' + C_2' + C_3'}, \qquad (22)$$

where

$$C_1' = \frac{\mu_2^v - r_{ve}(\mu_2^v \mu_2^e)^{1/2}}{\alpha^2 - \gamma_v^2}, \quad C_2' = \frac{\mu_2^e - r_{ve}(\mu_2^v \mu_2^e)^{1/2}}{\alpha^2 - \gamma_e^2}, \qquad (23)$$

$$C_3' = -\frac{1}{\alpha}\left[\frac{\gamma_v \mu_2^v}{\alpha^2 - \gamma_v^2} + \frac{\gamma_e \mu_2^e}{\alpha^2 - \gamma_e^2} - \frac{r_{ve}(\mu_2^v \mu_2^e)^{1/2}(\gamma_v + \gamma_e)(\alpha^2 - \gamma_v \gamma_e)}{(\alpha^2 - \gamma_v^2)(\alpha^2 - \gamma_e^2)}\right], \tag{24}$$

and $C_1' + C_2' + C_3' = \mu_2^h$. In particular, the autocorrelation coefficient r^h of the sea level is equal to:

$$r^h = r^h(1) = \frac{C_1' e^{-\gamma_v} + C_2' e^{-\gamma_e} + C_3' e^{-\alpha}}{C_1' + C_2' + C_3'}. \tag{25}$$

The cross-correlation function $r^{hv}(\tau)$ between the sea level and river inflow is equal to:

$$r^{hv}(\tau) = \begin{cases} \dfrac{e^{\gamma_v \tau}\left[\mu_2^v - r_{ve}(\mu_2^v \mu_2^e)^{1/2}\right]}{(\alpha + \gamma_v)(\mu_2^v \mu_2^e)}, & \tau < 0 \\[4mm] \dfrac{1}{(\mu_2^h)^{1/2}}\left[\dfrac{r_{ve}(\mu_2^e)^{1/2}}{(\alpha - \gamma_e)}A - \dfrac{(\mu_2^v)^{1/2}}{(\alpha - \gamma_v)}B\right], & \tau \geq 0 \end{cases} \tag{26}$$

where

$$A = e^{-\gamma_e \tau} - \frac{(\gamma_v + \gamma_e)e^{-\alpha\tau}}{(\alpha + \gamma_v)}, \qquad B = e^{-\gamma_e \tau} - \frac{2\gamma_v e^{-\alpha\tau}}{\alpha + \gamma_v}.$$

In a similar way the mutual correlation function between the sea level and evaporation can be obtained.

3. Test of the Applicability of the Model to the Caspian Sea

The model can be applied for the description of a certain hydrological system if the assumptions adopted when constructing the model are satisfied for this system. In the case of the Caspian Sea, it is necessary to check for the following assumptions:
- the linearity of the dependence between the Sea area F and the water level h;
- the relatively small degree of changes in F under the variations of the level h;
- the possibility of the linearization of the dependence between outflow from the Sea and the water level;
- the possibility of the simulation of river inflow and effective evaporation by the first-order autoregressive processes (AR(1)-processes).

The dependence between the area of the Caspian Sea and the Sea level is rather close to linear. The linear approximation of this dependence is:

$$F = 375 + 15h, \tag{27}$$

98

where F is measured in thousands km^2, the level h is counted from the level -28.0 m fixed as the beginning (Fig.2a). As the variations of sea level are in the range of approximately $\pm$ 2 m, the changes of the Sea area are insignificant in comparison with the mean area of the Sea, and thus the approximation (4) is satisfied.

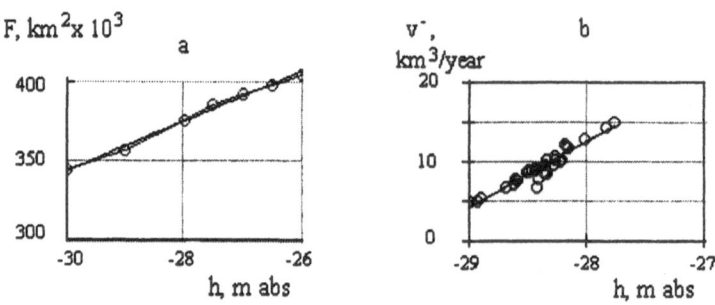

Figure 2. Area of the Caspian Sea (a) and outflow from the Sea into the
Kara Bogaz Gol Bay (b) *vs* the Sea level.
Circles denote observations and lines - linear approximations.

In general, the dependence between the outflow from the Caspian Sea into the Kara Bogaz Gol Bay and the Sea level is non-linear and for the various periods this dependence is not permanent. However, in the case when the variations of Sea level occur over a rather small range (for example, during 1946-1980, under the fixed hydraulic conditions in Kara Bogaz Gol Strait), this dependence can be approximated by a linear one [6]:

$$v^-(h) \approx 13.1 + 8.6h, \qquad (28)$$

where the level h is measured from the mark -28.0 m (Fig.2b).

The Caspian Sea level variations and the main components of the Sea water balance are shown in Fig.3 and Fig.4, respectively.

Usually, the river inflow into the Caspian Sea is simulated as an AR(1) process [2, 3, 17,19,20]. In spite of the fact that the natural mode of the river inflow into the Sea after the construction of the Volga-Kama reservoir cascade (1960s) is broken, this simulation is justified because the artificial changes in the long-term oscillations of the river inflow are not too large. These changes concern mainly the mean annual river inflow into the Sea. It is considered that due to irreversible withdrawals, the mean river inflow has decreased approximately by 30-40 km^3/year.

h, m abs

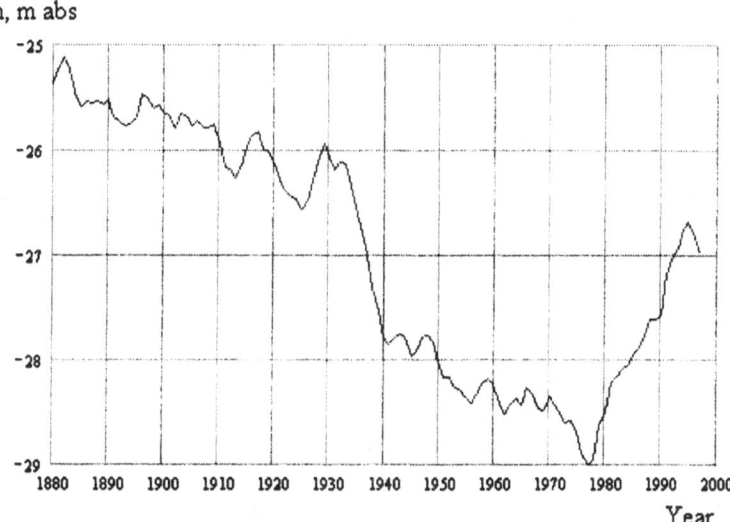

Figure 3. The Caspian Sea level variations during 1880-1997.

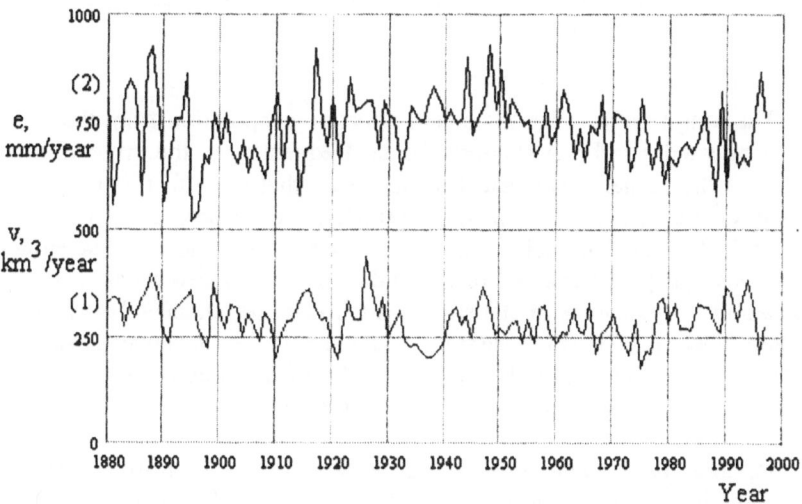

Figure 4. The river inflow into the Caspian Sea (1) and the effective
evaporation (2) from the Sea area.

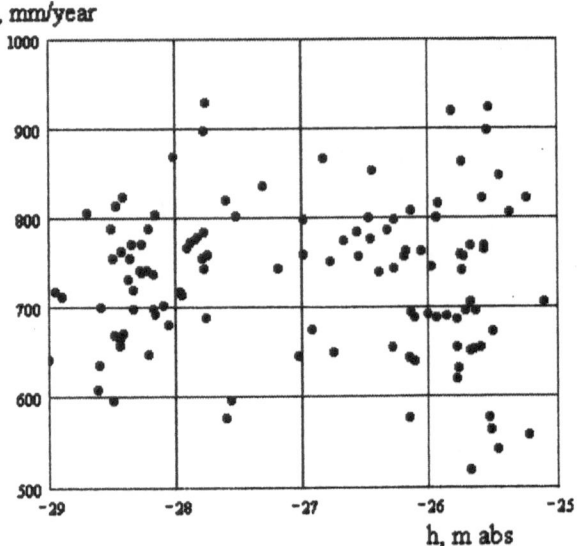

Figure 5. The effective evaporation from the Caspian Sea surface *vs* the Sea level.

The long-term oscillations of the effective evaporation from the Caspian Sea area are simulated by the AR(1)-process as frequently as the river inflow. In the framework of the model developed, it is necessary that the effective evaporation does not depend on the Sea level variations. Actually, the evaporation from the Sea area depends on the Sea level, especially in the shallow Northern Caspian Sea. However, this relation is weak [13]. The analysis of actual data does not allow reliable conclusions to be made about the existence of statistically significant dependence between levels and evaporation. There is no reason to assume a functional relation between evaporation and Sea level, as it is done in [17,19,20]. The model of the long-term oscillations of the effective evaporation is considered as an AR(1)-process.

The assumptions made when constructing the model are thus fulfilled. The dynamic-stochastic model of the hydrological system with two mutually correlated input processes (the river inflow and the effective evaporation) and output processes (the variations of the Caspian Sea level) is represented by a system of linear stochastic equations of 3rd order.

The variations of the Caspian Sea level in this model are described by the components of the three-dimensional Markov process. It is interesting to note that in general, the conditional expectation of the Caspian Sea level represents a non-monotonic function of time. Some examples are shown in Fig.6a.

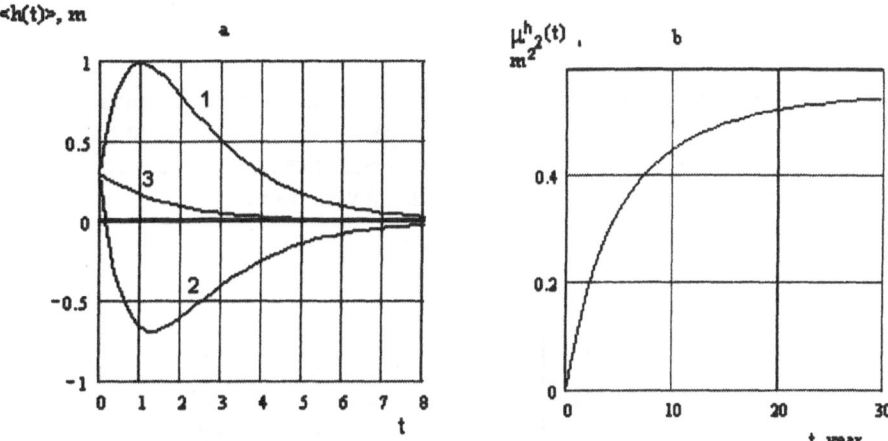

Figure 6. Examples of the conditional mathematical expectation of the open lake level under different
initial conditions (a) and the conditional variance of the Caspian Sea level (b).
1) $v_0 = 2$, $e_0 = -2$; 2) $v_0 = -2$, $e_0 = 2$; 3) $v_0 = 0$, $e_0 = 0$; $h_0 = 0.3$, $\alpha = 0.6$, $r_v = r_e = 0.3$.

The non-monotonic character of the conditional expectation of the level $<h(t)>$
as a function of time is not exclusive to the Caspian Sea [5, 8]. The conditional variance
of the Caspian Sea level is a monotonically increasing function of time (Fig.6b).

4. Application of the Dynamic-Stochastic Model for the Evaluation of the Changes in the Long-term Variations of the Caspian Sea Level

The analytical expressions obtained in Section 2 represent the statistical characteristics
of the sea level as a function of the statistical parameters of the river inflow and the
effective evaporation, and the morphometric and hydraulic parameters of the lake. The
changes in these parameters effect the lake level characteristics. Hence, it is possible to
evaluate the changes in level induced by changes in the input processes and the lake
parameters.

The outflow parameter of the Caspian Sea characterizes the hydraulic
conditions in the strait connecting the Sea and the Kara Bogaz Gol Bay. Let us evaluate
the impact of the changes in this parameter in the modern time on the Sea level
variations. The Kara Bogaz Gol Bay is located on the Turkmenian coast of the Caspian
Sea. It is a unique geographical object, being the largest salt container in the world. The
area of the Bay is approximately 20 000 km² at the Bay water level of -27 m, which is
of the same order of magnitude as the area of Ladoga Lake or Lake Ontario. The main
incoming part of the Bay water balance is the marine water from the Caspian Sea,
which flows into the Bay through the Kara Bogaz Gol Strait with a length of 10-12 km.
The water current in the Strait is always directed from the Sea to the Bay due to the
significant evaporation from the surface of the Bay. Under normal hydraulic conditions

in the Kara Bogaz Gol Strait, the higher the Sea level the more the marine water flows into the Bay.

The evaporation from the Bay area is the sole outgoing component of the Bay water balance. The Bay evaporates 16-20 km^3 / year at the water level in the Bay close to the -27 m mark. With respect to the Caspian Sea, the Kara Bogaz Gol Bay is the natural regulator reducing the range of the Sea level variations.

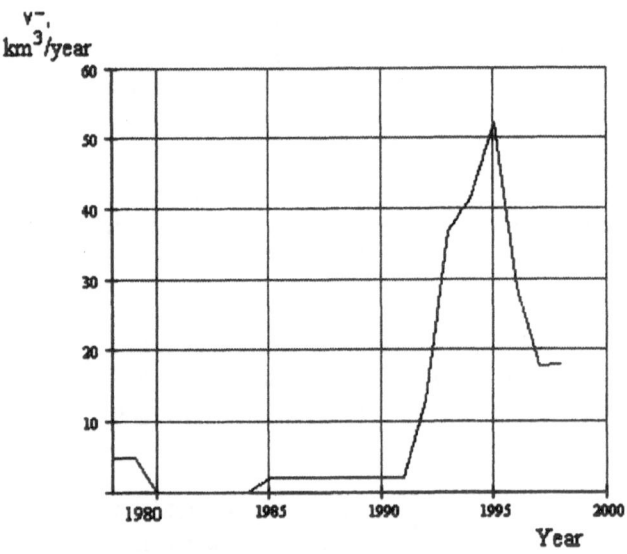

Figure 7. The outflow v from the Caspian Sea into the Kara Bogaz Gol Bay during 1978-1997.

The operation of this natural regulator was terminated in 1980. The decrease of the Caspian Sea level was the main reason for the dam construction, which terminated the outflow of the marine water into the Bay. In 1977 the Sea level reached the mark of -29 m. This level is critical for the conservation of sturgeon, which is the main biological resource of the Caspian Sea. At sea levels lower then -29 m, a sharp decrease of the bioproductive areas in the shallow waters of the Northern Caspian Sea occurs.

The termination of the marine water outflow from the Sea into the Bay took place just at the beginning of the rise of Caspian Sea level. By 1992 the Sea level had risen almost by 2 m, up to a mark of -26.9 m. The contribution of the Kara Bogaz Gol Bay cut-off to this increase of the Sea of level is estimated to be approximately 0.3-0.4 m.

In June 1992, by the decision of Mr. Niyazov, the President of the Republic of Turkmenistan, the dam was destroyed. At that time, the difference between the water levels in the Caspian Sea and in the Kara Bogaz Bay was approximately 7 m. The marine water gushed into the Bay. and the sandy shores of the Strait connecting the Sea and the Bay were eroded. The cross-sectional area of the Strait doubled according to the data presented by the Turkmenistan Hydrometeorological Service. A huge volume of marine water entered the Bay, reaching a maximum of 52 km^3/ year (Fig.7). Note, that before the cut-off, the outflow volumes were only 5-10 km^3 / year. The dynamics of the filling of the Bay is shown in Fig.1 (right).

At present, as a result of the filling up of the Bay and the reduction of the difference between the water levels in the Sea and the Bay, the volumes of marine water flowing into the Bay has decreased approximately to 17-20 km^3/year.

After the dam destruction, the degree of influence of this outflow upon the Sea level variations has varied. Our studies [6] have shown that all basic characteristics of the Caspian Sea level variations functionally depend on the outflow mode (Fig.8). There are two main characteristics of the outflow mode, namely, the average of the outflow volumes and the outflow parameter. It is possible to describe the influences of these characteristics on the Sea level variations in a simplified manner as follows.

Most obvious (but not unique) is the impact of the changes in the average of the outflow volumes on the Sea equilibrium level. The higher the average of the outflow, the lower is the equilibrium mark of the level. Roughly speaking, the higher the average outflow volumes, the lower is the average level of the water in the Sea. The outflow parameter influences to the range of the Sea level variations and, consequently, the level variance. The maximum range of the level variations would occur (with other characteristics being the same) at zero outflow, i.e. at the cut off the Bay.

The equilibrium level marks for the Caspian Sea differ considerably for the periods before and after the dam destruction, being -26 m and -28 m respectively. Thus, restoration of the marine water outflow decreases the equilibrium level by approximately 2 metres. The increase in the outflow parameter from zero (when the Bay was cut off) up to 0.06-0.08 $year^{-1}$ (after the dam destruction) resulted in a decrease of the standard deviation of the Sea level from 1 m to 0.5 m (Fig.8). Undoubtedly, these significant changes in the characteristics of the Sea level variations should be taken into account when assessing the applied problems related to the Caspian Sea level.

From the considerations above, it is obvious that the influence of the outflow from the Sea on the Caspian Sea level variations is rather essential. After destruction of the dam the Kara Bogaz Gol Bay has absorbed the volume of the Caspian Sea, which is approximately equivalent to a layer of 0.5 m thickness. If the dam had not been destroyed, the Sea level would have risen up to -26 m (Fig.9, curve 1).

It is interesting to compare the actual level of the Caspian Sea from 1978-1997 with the hypothetical scenarios of the Sea level variations under the various modes of outflow from the Sea into the Bay (Fig.9).

104

If the dam in Kara Bogaz Gol Strait had not been constructed, the dependence between the outflow from the Sea into the Bay and the Sea levels would correspond to the one that existed from 1946-1978. In this case, during 1978-1997 the maximum of the Sea level would have been below the actual marks (Fig.9, curve 2) by approximately 0.3 m (Fig.9, curve 3). Thus, the construction of the dam had a negative consequence, namely, an additional rise of the Sea level. It caused economic loss due to the submerging of the coastal areas.

During 1978-1997, the Sea level rise would have been minimal under the modern outflow mode formed after 1992 (Fig.9, curve 4).

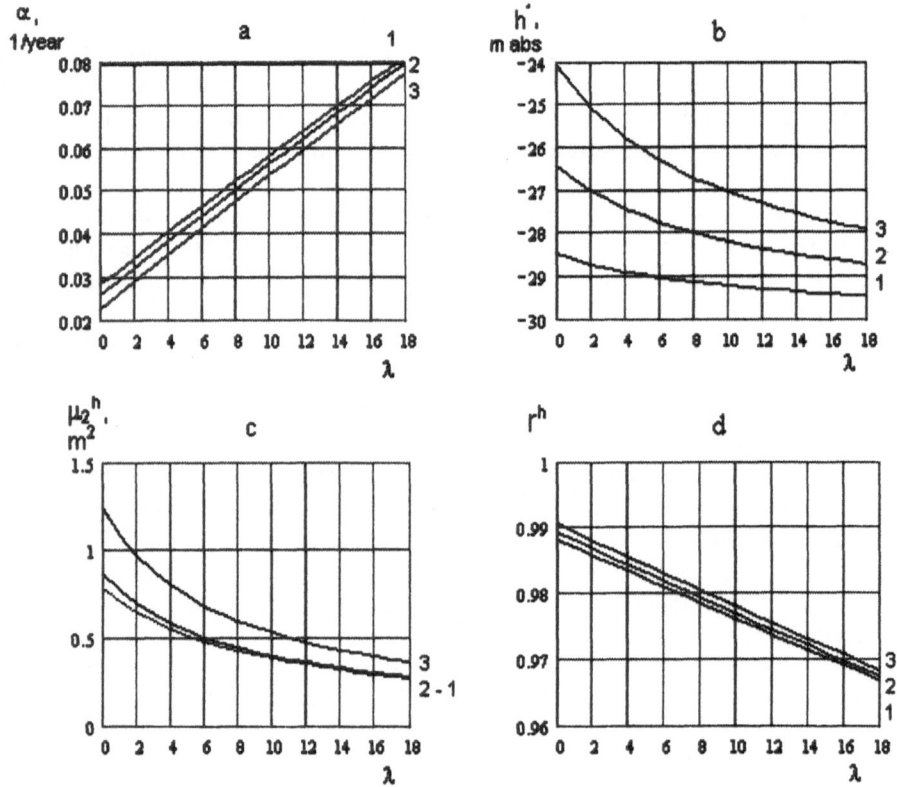

Figure 8. The Caspian Sea: (a) - inertial parameter α; (b) - equilibrium level h *; (c) - variance of the Sea level μ_2^h; (d) -autocorrelation coefficient r^h for some variants of the Sea water balance: 1) $<v^+> = 275$ km^3/year, $<e> = 0.75$m/year; 2) $<v^+> = 290$ km^3/year, $<e> = 0.73$ m/year; 3) $<v^3> = 300$ km^3/year, $<e> = 0.70$ m/year.

h, m abs

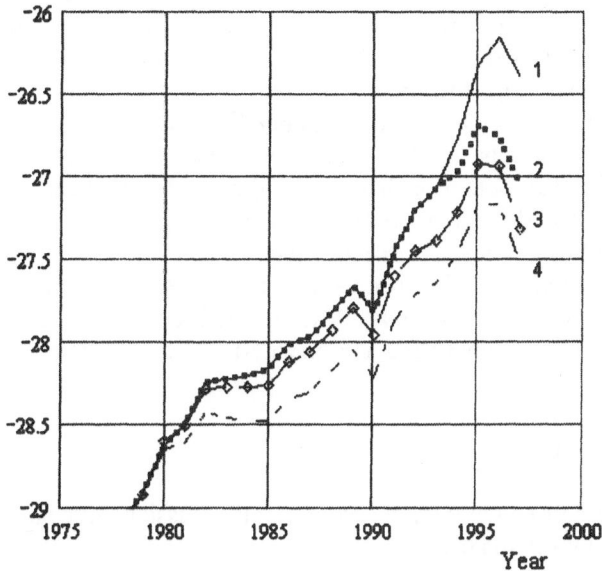

Figure 9. The Caspian Sea level variations under some modes of the outflow v-
from the Sea into the Kara Bogaz Gol Bay.
1 - on condition that v ˙ = 0, 2 - actual data, 3 - on condition that v˙ (h)
corresponds to the hydraulic mode that existed in Kara Bogaz Gol Strait
during 1946-1980, 4 - on condition that v˙ (t) corresponds to the new
outflow mode formed after destruction of the dam in 1992.

The difference between the maximum level marks for scenario 1 (absence of
the outflow from the Sea) and 4 (modern mode of outflow) is equal to approximately 1
m, which is quite essential. It is possible to state that in this case the Sea level rise
would not have been catastrophic. The increasing of the governing influence of the
outflow from the Sea into the Bay can be considered to be a positive result of the dam
destruction.

Thus, the consequences of the experiments (i.e. construction and destruction of
the dam) with the regulation of the outflow into the Kara Bogaz Bay to control the level
of the Caspian Sea are ambiguous. The additional increase in Sea level after the
construction of the dam is negative. The growth of the influence of the outflow from the
Sea into the Bay on the Sea level variations can be evaluated as positive.

5. Conclusions

A physically-based model of a stochastic reservoir with a two-dimensional input process is proposed. It is shown that this model can be used for simulating the long-term variations of the levels of the Caspian Sea.

The model developed can be applied for forecasting the Caspian Sea level fluctuations. It can also be used to assess the impact of climatic and anthropogenic changes in the Caspian Sea water balance on the Sea level variations. For example, it has been shown that the Sea level statistical characteristics depend on the mode of outflow from the Sea into Kara-Bogaz-Gol Bay.

The developed model generalizes the "linear" models of the stochastic reservoir suggested by other authors. A further refinement of the model envisages a simulation of the non-stationary and non-linear effects in the Sea level variations.

6. Acknowledgement

The research was carried out with the support of the Russian Foundation for the Basic Researches (grant 99-05-64361).

7. References

1. Babaev, A.G. (1997) Ecological, Social and Economic Problems of the Turkmen Caspian Zone, in M.H.Glantz and I.S.Zonn (eds.), *Scientific, Environmental and Political Issues in Circum-Caspian Region*, Kluwer Academic Publishers, Dordrecht, pp.97-104.
2. Bagrov, N.A. (1961) On the variations in levels of closed lakes, *Meteorologia i Gidrologia* **6**, 41-48 (in Russian).
3. Budyko, M.I. and Yudin, M.I. (1960) On variations in the levels of non-terminal lakes, *Meteorologia i Gidrologia* **8**, 15-19 (in Russian).
4. Frolov, A.V. (1985) *Dynamic-Stochastical Models of the Long-Term Variations in the Non-Terminal Lake Levels*, Nauka, Moscow (in Russian).
5. Frolov, A.V. (1989) Effects of autocorrelation in river inflow and effective evaporation on the regime of lake level variations, *Meteorologia i Gidrologia*, **4**, 94-101 (in Russian).
6. Frolov, A.V. (1998) Influence of resumption of the marine water outflow into the Kara Bogaz Gol Bay on the long-term sea level fluctuations in the Caspian Sea, *Meteorologia i hydrologia* **7**, 87-97 (in Russian).
7. Gates, D.G., and Diesendorf, M. (1977) On the fluctuations in levels of closed lakes, *Journal of Hydrology* **33**, 267-285.
8. Golitsyn, G.S., Ratcovich, D.Ya., Fortus, M.I. and Frolov, A.V. (1998) On the modern rising of the Caspian Sea level, *Wodnye resursy* **25**, 133-139 (in Russian).
9. Kaczmarek, Z. (1971) Some problems of stochastic storage with correlated inflow, *IASH Publications* **100**, 435-439.
10. Klemeš, V. (1974) Probability distribution of outflow in a linear reservoir, *Journal of Hydrology* **21**, 395-414.
11. Klemeš, V. (1978) Physically based stochastic hydrological analysis, *Advances in Hydroscience* **11**, 285-356.

12. Kritsky, S.N.and Menkel, M.F (1946) Some positions of the statistical theory of the water bodies levels variations and its application to the study of the Caspian Sea, *Proceedings of the First Conference on the River Flow Regulation*, USSR Academy of Sciences, Moscow, 76-93 (in Russian).

13. Kritsky, S.N., Korenistov, D.V. and Ratcovich D.Ya. (1975) *The variations in the Caspian Sea level*, Nauka, Moscow.

14. Lloyd, E.H. (1963) Reservoirs with serially correlated inflows, *Technometrics* **5**, 85-93.

15. Lloyd, E.H. (1993) The stochastic reservoir: exact and approximate evaluations of storage distribution, *Journal of Hydrology* **151**, 65-107.

16. Muzylev, S.V. (1980) Theoretical-probabilistic analysis of the variations in the terminal lakes levels, *Wodnye Resursy* **5**, 21-40 (in Russian).

17. Muzylev, S.V., Privalsky ,V.E. and Ratcovich, D.Ya. (1982) *Stochastic Models in Engineering Hydrology*, Nauka, Moscow (in Russian).

18. Phatarfod, R.M. (1979) The bottomless dam, *Journal of Hydrology* **40**, 337-363.

19. Privalsky, V.E. (1988) Modeling long term lake variations by physically based stochastic dynamic models, *Stochastic Hydrology and Hydraulics*, **2**, 303-315.

20. Ratcovich, D. Ya.(1993) *Hydrological Principles of Water Supply*, Water Problems Institute of the Russian Academy of Sciences, Moscow (in Russian).

GEOPHYSICAL TIME SERIES AND CLIMATIC CHANGE

A sceptic's view

V. KLEMEŠ
Consultant in Water Resources and Hydrology
3460 Fulton Road, Victoria BC, V9C 3N2, Canada

1. Preface

In these days, the notion of 'climate change' has a distinctly negative and passive connotation. It raises concerns and worries, it is something that should be avoided, prevented, or at least minimized. It is seen as an inadvertent byproduct of the technology-dominated civilization, as the ultimate 'pollution' and we see ourselves as its victims. This was not always so. Just half a century ago, man was seen as the future master of climate. The connotation of climate change was 'control' rather than 'inadvertent byproduct'. Climate change was the aim, was seen as an instrument for bettering the human condition, and sometimes was exalted as the ultimate triumph of technology and the human genius.

From modest experiments with cloud seeding in Colorado, to Joseph Stalin's grandiose plans for the 'transformation of nature' by the 'great constructions of communism' - that was the 'climate of climatic change' fifty years ago.

Indeed, were it not for that 'climate', I would not be talking here today. In the 1950 Czechoslovakia, the best chance to get university education for those with 'politically incorrect' backgrounds was to apply for admission into Civil Engineering, thereby demonstrating a willingness to 'contribute to the socialist construction' and, in the process, develop the 'socialist consciousness'. And so I too was admitted and soon was marching, in countless 'voluntary' student brigades, with a shovel on my shoulder, to the tune

> *"The rain shall we command*
> *And the river flow,*
> *And order the wind when*
> *And where to blow..."*

No doubt that the climate has changed since those days. It still does.

M.V. Bolgov et al. (eds.), Hydrological Models for Environmental Management, 109–128.
© 2002 *Canadian Water Resources Association. Printed in the Netherlands.*

2. Introduction

The title of this lecture makes allusion to three different notions: *'time series'* has a mathematical connotation, *'geophysical'* suggests a connection with some signals emanating from this earth, and *'climatic change'* points to a concern about an apparent nonstationarity of the present condition. Indeed, the focus of this lecture can be expressed by the following question:

What can mathematics tell us about the physical signals in our nonstationary world?

My answer to this question is

Little!

and, by looking closer at the three above components of the question, I shall try to explain why.

3. Mathematics

In regard of mathematics, I will start by quoting what Bertrand Russell said at the opportunity of his 80th birthday (cited from Kempthorne [1]):

> *"I set out with a more or less religious belief in a Platonic eternal world, in which mathematics shone with a beauty like that of the last cantos of the Paradiso. I came to the conclusion that the eternal world is trivial and that mathematics is only the art of saying the same thing in different words."*

Russell has rendered us an invaluable service by committing this insight to paper: We do not have to wait until the age of 80 to realize the same thing and can devote some of the time saved to ponder the implications of his profound statement. What I think he meant was to emphasize that, in dealing with mathematics, we deal with a world of our own creation, a world of ideas which we can manipulate to our liking using rules which we ourselves have imposed, that mathematics does not create new knowledge about the world and can only express our ideas about it. Mathematics is innocent of the value of the equations by which we describe these ideas - it is us who must be blamed when they are wrong, and can take credit when they turn out to be right. Mathematics is like language and grammar: it can help us formulate what we know, and ask questions about, or hide, what we do not - but the use of correct grammar offers no clue to the value of our statements, nor does it even guarantee that they make any sense at all.

Consequently, our mathematical models, including time-series models, by which we try to describe geophysical records are only as good as is our understanding of the processes that generated them. Many famous statisticians have emphasized this fundamental fact in the context of statistical analysis, time series, and probabilistic modelling in general. I have collected a number of pertinent quotations elsewhere [2] and will repeat only two of them here. Thus M. S. Bartlett once said that "unless the statistician has a well defined and realistic [causal] model of the actual process he is studying, his analysis is likely to be abortive"; and on the topic of time series analysis we may note the following observation of Kendall and Stuart: "The ultimate object of analysis of time series - as of statistical analysis as a whole - is to arrive at a deeper understanding of the causal mechanisms which generated it."

To put it in a nutshell, a geophysical process must first be analyzed and understood from the physical point of view before it can be adequately mathematically described. Its time-series model should be a synthesis of what we know about the prototype, and a proper synthesis cannot come before a proper analysis - and analysis, including statistical (and time-series) analysis, means **asking questions** as Kempthorne [1] admonished his fellows statisticians more than a quarter of a century ago. How many of them did get the point, I do not know; what I do know is that not many stochastic hydrologists did - they still seem to prefer offering answers without first asking questions; that is to say, they concentrate on purely descriptive empirical models. But, as I have tried to explain on many occasions before [3,4,5], while such models can be useful for engineering and other applications, they contribute nothing to advancing the understanding of the processes involved and are thus of no value for making inferences about such matters as climate change.

These empirical time-series models are typically based on one or more of the following principles: (1) fitting of the graphical and numerical patterns of an observation record, (2) forcing an observation record upon the Procrustean bed of one's pet model, and (3) prejudging, based on one's own biases and prejudices, the results of analyses not done, second-guessing the causes of the observed patterns and, as a result, adopting wrong model structures.

To kill three birds with one stone, so to speak, I shall use one rather old Canadian paper [6] which illustrates all these three approaches simultaneously. It is concerned with the development of a time-series model for the annual flows of the Nile River at Aswan, based on the historical flow record for the period 1870-1945 as shown in Fig.1. The pattern of this plot, in particular the conspicuous downward trend in its first half, **raises the legitimate question** about its causes and one would think that a serious attempt at their clarification would be the logical first step in the search for an appropriate structure of the model. Instead, the authors considered only one possibility: the first filling of the old Aswan dam which occurred in the season 1902-1903. And they did not take it as a hypothesis to be verified but as an established fact - the one fact on which the entire structure of their model rests. Why? One can only guess and my guess is as follows:

Firstly, the authors' overall aim was to **model a pattern**, not to understand its origin. Secondly, their specific aim was to push (in their words, "demonstrate the potential" of) their **pet model** - 'intervention analysis' - which is, by definition, a modelling of an abrupt change in a process. So, the apparently smooth decline in the central tendency suggested by the pattern of the plot had to be forced onto the Procrustean bed of a 'sudden jump' in the mean. Thirdly, feeling a need to justify this 'jump' but being in no mood to analyze its origin seriously, the discrete event of the Aswan dam's filling seemed like a self-evident cause, especially to engineers which is what most of the authors of that paper were - an example of **jumping to conclusions** about the 'underlying physics', based on personal biases rather than objective analysis.

The essence of the result of what the authors claimed to be a 'rigorous analysis' is the representation of the historic flow record as a time series fluctuating about a mean which undergoes a downward jump of 22% exactly at the time when the Aswan dam was filled (Fig.1). In the paper, this is described by two elegantly looking transfer function equations which the authors claim "are stochastic models that can be used for forecasting and simulation of the average annual flows of the Nile River at Aswan". It is true that, for simulations of 'statistically similar' time-series for the underlying historical period, they can be used (though it is difficult to think of any

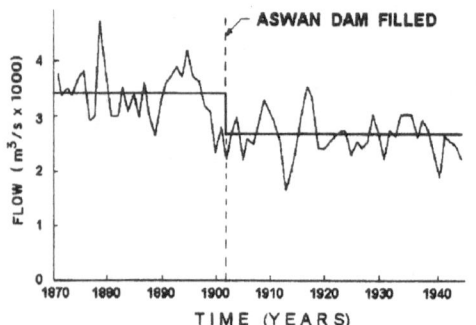

Figure 1. Annual flows of the Nile River at Aswan and an 'intervention model' of their mean [6].

meaningful purpose or use of such simulations themselves), but if used for forecasting beyond that period the result would, to use Bartlett's words, be abortive because it is not based on a realistic model of the actual process.

The lack of realism of the authors' postulated causal model (the filling of Aswan dam) for the sudden drop in the mean flow of the Nile is exposed by the fact that, around the turn of the century, a similar drop in the mean also occurred in precipitation over a vast region including much of the tropical Africa and the West Indies (Fig.2a,b,c) (Fig.2 has been adapted from Todini and O'Connell [7]) - a drop that can hardly be attributed to the Aswan dam! Indeed, Fig. 2d

113

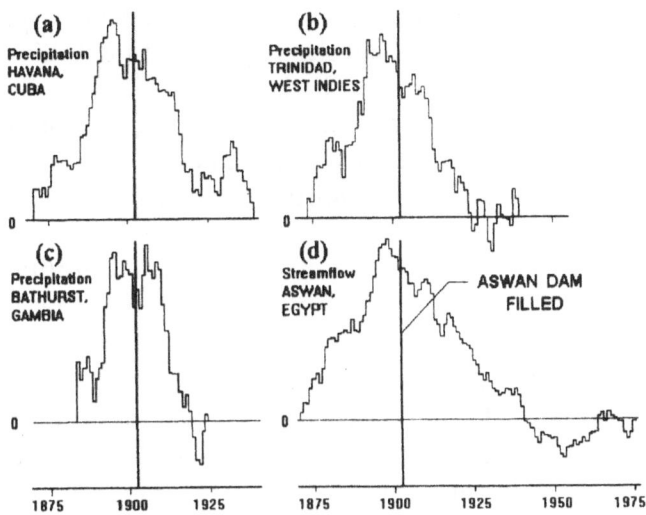

Figure 2. Residual mass-curves of three tropical rainfall records and Nile River flows at Aswan [7].

indicates that the drop in the Nile flows commenced already before 1900 as did that in the tropical precipitation - which in fact is also suggested by Fig.1 but was ignored because it did not fit into the preconceived aim of fitting an 'intervention' model to the data. Fig.2d also graphically illustrates the 'abortive' forecasting result of this 'rigorous' model: the effect of the 'Aswan dam intervention' seems to have magically vanished after 1950, just about five years after the end of the historical record used for the development of the model.

Apart from the dangers discussed above, there is one important additional but often overlooked fact which I shall call the 'mischief factor' of mathematics. It sometimes causes mathematics to frustrate rather than facilitate a scientific discovery: The specific mathematical method used in data analysis may introduce into the result some features which are then wrongly attributed to the physical process being analyzed. Probably the most common case arises in correlation analysis where the mischief factor often manifests itself as spurious correlation, but less known mischief factors lurk in other methods, e.g. moving averages, L-moments, mass-curves, etc.

As an illustration, I will discuss the mischief factor inherent in the 'residual mass-curve'. Such mass-curves are commonly used to represent precipitation and streamflow records (see Fig.2) and their patterns are often regarded as manifestations of climate fluctuations and changes. This makes their mischief factor particularly relevant to the topic of this lecture. Residual mass-curve of a series (observed record) x_i, $i=1,2,...,n$, is the transformed sequence y_i, $i=1,2,...,n$, defined as the cumulative sum of the departures of x_i from the sample mean $<x>_n$, that is

$$y_i = \sum_{j=1}^{j=i} (x_j - <x>_n) .$$ (1)

Residual mass-curves invariably exhibit a more-or-less pronounced cyclicity which some authors have regarded as a conclusive proof of cyclicity in

114

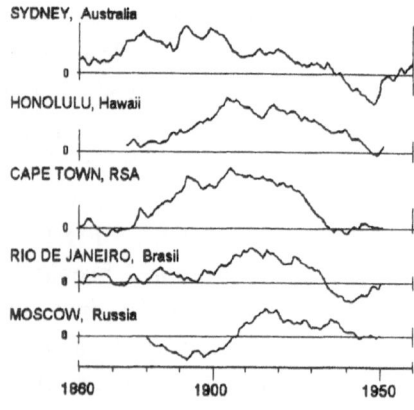

Figure 3. Residual mass-curves of some precipitation records.

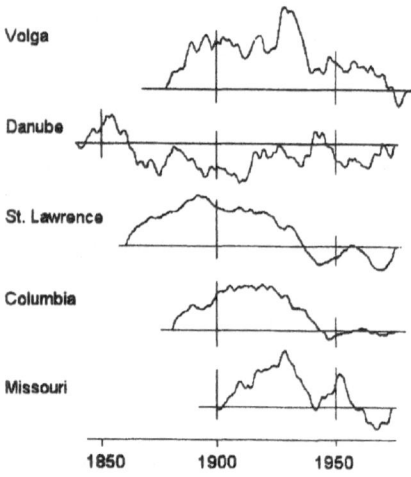

Figure 4. Residual mass-curves of some streamflow records.

hydrological processes. For example, such were the conclusions reached by Williams [8] on the basis of mass-curve plots reproduced in part in Fig.3, and by Shiklomanov and Markova [9] on the basis of similar plots some of which appear in Fig.4.

Such claims may sometimes be justified, but sometimes they may be based on the 'mischief factor' which causes a repeated application of eq.(1) to produce a cyclic-looking sequence from an original sequence of almost any kind in about three steps (Fig.5). Moreover, this sine wave always extends over the whole sample, regardless of its size **N**, as shown in Fig.6 (Figs. 5 and 6 are adapted from [10]).

This discovery had made me quite proud but my son, who happens to be a mathematician, explained to me that it was mathematically trivial, that simply all but the first Fourier component of the series were gradually eliminated by the repeated integrations. I protested that, while it could be trivial in the 'Platonic eternal world' of mathematics, it was far from trivial in the real world where it can, and sometimes does, lead to wrong climatic, hydrologic, economic and other inferences, and we wrote a joint paper about it. Eventually, even the editor of the prestigious *Water Resources Research* was converted to my view and our paper was published [10].

Another aspect of the mass-curve 'mischief factor' is that the apparent synchronicity - or a lack of it - of the phases of mass-curve plots (used, for example, by Lalykin [11] to propose a regionalization scheme of the territory of the former USSR on the basis of 'co-phased' river runoff [Fig.7]) can sometimes be a fiction created by comparing plots for series of different lengths (as Lalykin did). This is demonstrated in Fig.8 where, due to the different sample means for the periods T_1 and T_2, the two corresponding residual mass-curves of the time-series of the process x can have opposite phases in the same epoch Δt. It certainly would be quite extraordinary if a physical process were not in phase with itself during some periods of time!

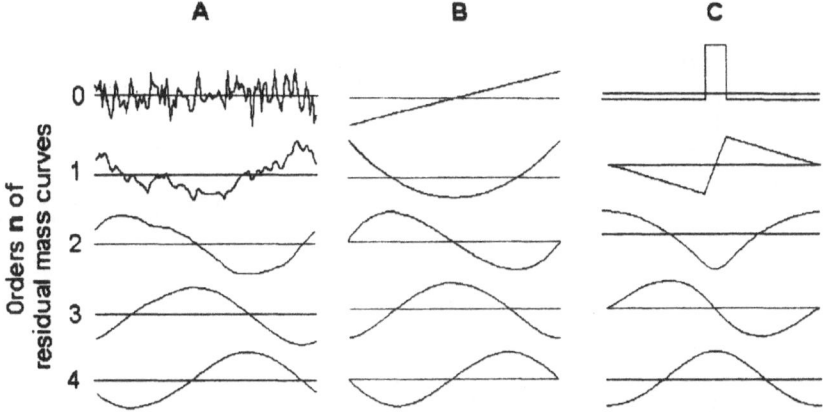

Figure 5. Examples of residual mass-curves of higher orders for (A) streamflow record, (B) linear trend line, (C) discrete pulse[10].

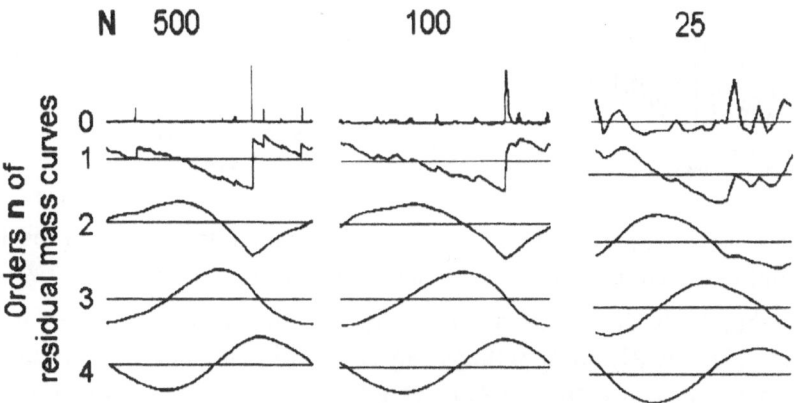

Figure 6. Examples of residual mass-curves of higher orders for different-size samples from the same parent random series [10].

Such and similar effects may contaminate interpretations of past climatic trends, in particular when they are reconstructed from proxy data.

4. Nonstationarity

In the framework of mathematical theory of stochastic processes, concepts of stationarity and nonstationarity can be defined quite unambiguously (though in different ways) and this carries over into time-series models. However, I am led to believe that in the context of thereal world these concepts are - like beauty and ugliness - in the eye of the beholder, in the horizon the eye can reach. To put it more technically, they are a matter of time scale. Fig.9 illustrates the point. It shows the global temperature record reconstructed from different sources and on different time scales [12]. If our horizon extends, say, just over the past 250 years (Fig.9D), we are tempted to say that the process is stationary; if it stretches

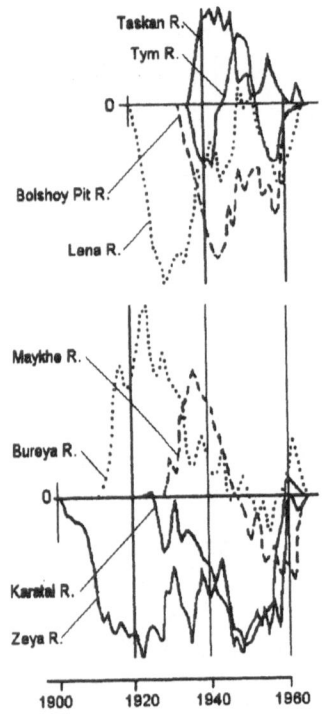

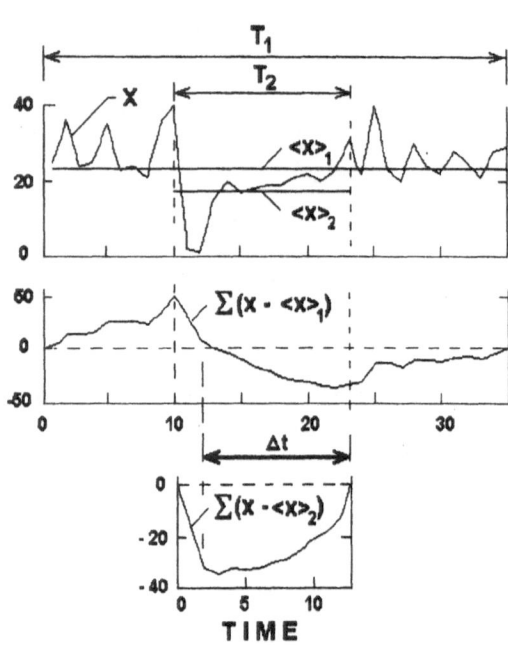

Figure 7. Residual mass-curves of
some streamflow records [11].

Figure 8. Illustration dependence of residual
mass-curve phase on the sample size.

over tens-to-hundreds thousands of years the last glacial period makes it look nonstationary (Fig. 9C,B) but it perhaps can be regarded as stationary again if we expand our window over a million years (Fig.9A). If we extend it still further we may find that the temperature process evolved through a nonstationary phase at the early age of the Earth, but then the Earth history itself may be just a 'fluctuation' in a billions-years long 'stationary' stage of the history of the universe, in turn preceded - as both the physicists and the theologians tell us - by a marked non-stationary period lasting anywhere from a few milliseconds to six days. Actually, some physicists are currently one step ahead of the theologians: they allude to (presumably stationary?) 'many-worlds' theories, 'tunnelling', etc., extending the time window beyond the beginning of time itself ...

Obviously, in the context of 'climatic change', we are not concerned with the 'ultimate answer' to the question of world's stationarity or otherwise, but with the **present trend** of the climate and its development in the several immediate future decades, with its 'dynamics' rather than with the nature of its 'time series'. Hopefully, in our search to determine it, we can learn from the historic realizations of geophysical processes and combine this knowledge with that of our 'anthropogenic forcing' to get an idea about the future course of events.

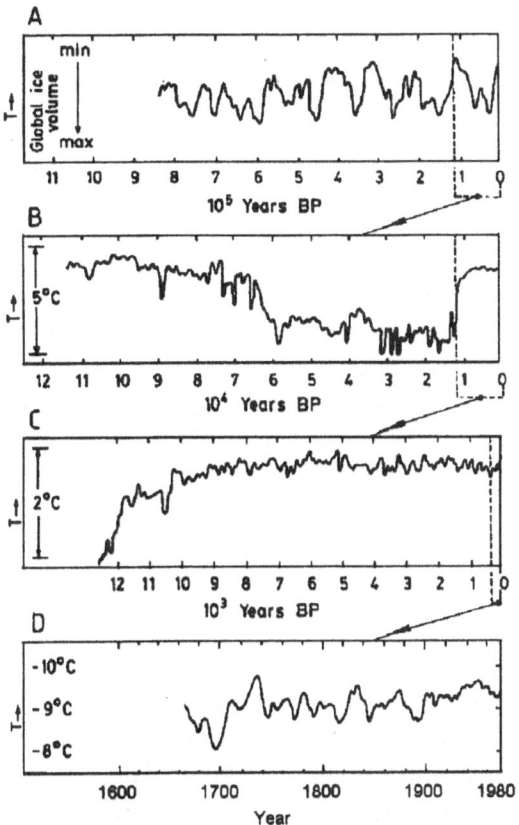

Figure 9. Global temperature fluctuations on different time scales [12].

In trying to grasp the present dynamics of the climate, I consider it rather irrelevant whether the geophysical records can be better described by stationary or by nonstationary time-series models, fractional-noise models or Box-Jenkins models, etc. - though I have no doubt that wars over these matters will be fought in the academia for years to come. Nor do I see any point in constructing time-series models for 'scenarios' of runoff, precipitation, temperature, etc., for the 'doubling of CO_2', 'year 2050', etc., all implying an onset of some 'stationary regime' in which it is hard to believe given the current rate of population growth, deforestation, urbanization and, above all, the modern economists' and politicians' postulate of 'GROWTH' as the supreme criterion of success of our civilization.

More meaningful is the quest of scientists to disentangle the interplay of the various component processes of the climate as reflected in the available data, and formulate **testable hypotheses (questions) about the expected transient climate behaviour** based on the present knowledge of its dynamics and anchored to its present state as the concrete initial and boundary conditions. This I would call an 'honest and humble search for signals' as opposed to the boastful claims of assorted 'modellers' about all kinds of 'climate-change effects', often based more on ignorance than on knowledge, motivated more by politics than by science, and reflecting prejudices rather than facts. Whether the latter tendencies can be suppressed is by no means certain because they are not limited to the present climate-change research. For, if Confucius can be trusted, they have been part and parcel of the intellectual climate for at least the last twenty five centuries (an unfortunate kind of 'climate stationarity', one might say). Or could it be that Confucius was referring to the present-day 'modellers' when he said (in 'The Great Learning')

"... their thinking is insincere because their wishes discolour the facts and determine their conclusions, instead of seeking to extend their knowledge to the utmost by impartially investigating the nature of things..." ?

5. Signals

The fact that signals (=imprints of specific mechanisms) are present in geophysical time series is rather unfortunate and complicates matters quite considerably. Just imagine how idyllic the world would be, how our stochastic models of time-series would work, if everything were random - not to say Gaussian! - noise! But this was not to be so. The harsh reality is that the 'Kisiel prayer' of the stochastic hydrologist [13],

"Oh, Lord, please, make the world linear and Gaussian!"

was not granted.

There are two problems: signals must first be detected and then they must be linked by meaningful, i.e. causal, relationships. In both these tasks, mathematics can be of great help but it also can deceive since it can be used 'to say in different words' what Confucius said in his words quoted above. This danger cannot be emphasized often enough since our interpretations can be affected by our own life experiences, professional and cultural biases and prejudices to a greater extent than we may wish to admit [14]. They inadvertently project themselves onto the patterns that we see before us and it is all too easy to give a mathematical form to the distorted images thus created. The case of the 'Aswan dam intervention' into the Nile flow record was one example. To emphasize the point, I will use one more example from my own experience [15].

Fig. 10a shows a plot of net monthly inflow (abscissa) versus runoff for the Rainy Lake Basin in Canada. In the absence of this specific knowledge and disregarding the units and the numerical labelling of the axes, what interpretation does the pattern alone suggest? A statistician could perhaps see in it a random sample from a bivariate Gamma distribution, an astronomer a cluster of galaxies, an etymologist a relation between abundances of honey bees and stingless bees (see later), while the Good Soldier Švejk would undoubtedly identify it as fly stains on the old picture of His Highness Emperor Frantz Joseph der Erste hanging in the Prague pub 'U kalicha' (Fig. 10b), the main source of Švejk's life experiences. Only after a painstaking 'impartial investigation of the nature of things' the true picture, the true signals, start to emerge: a signature of complex processes involving hydraulic reservoir routing as shown in Fig. 11, thermally and radiatively forced snowmelt and evapotranspiration effects, all locked in

a)   **b)**

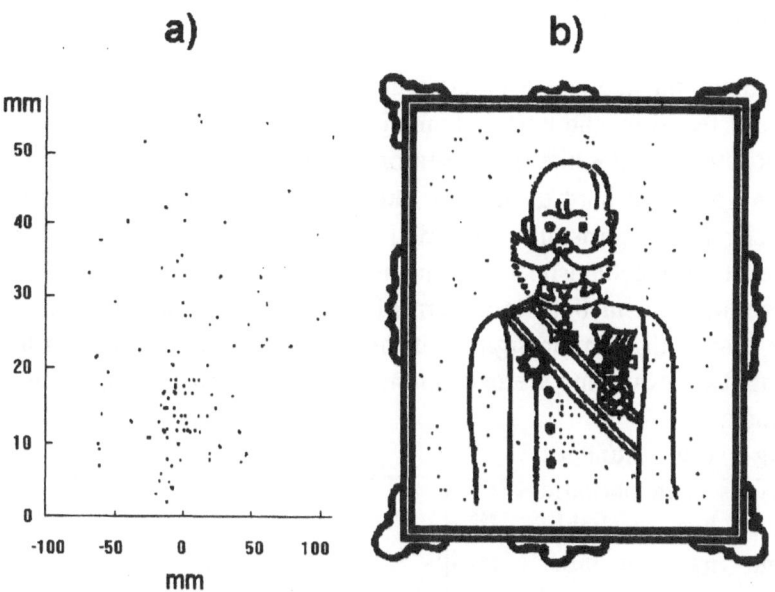

Figure 10. a) Net monthly inflow (abscissa) and runoff for Rainy Lake Basin, Canada [15]; b) A likely interpretation of plot (a) by the Good Soldier Švejk [14].

through intricate feedbacks, subject to various thresholds, changing boundary conditions, etc., etc.

The signals shaping the form of geophysical processes are many. In the following, I will therefore concentrate on only two kinds of them, those produced by two mechanisms which are ubiquitous in the physical world and of particular importance in climate formation. They are **rotation** (pulsation, oscillation) on the one hand, and **accumulation** in (and release from) storage on the other. Indeed, if God is a mathematician as some sources claim, his pet areas must be Fourier Analysis and Integral Calculus!

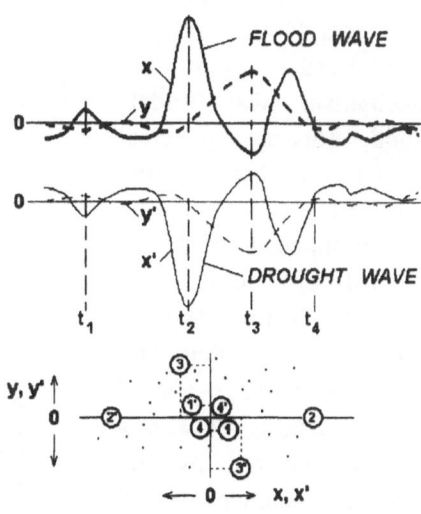

Figure 11. Typical relationship between inputs (x, x') and outputs (y, y') of a system of reservoirs.

120

5.1 ROTATION

Things rotate, pulsate and oscillate wherever we look: from neutrinos to eddies in a cup of tea, from the beat of human hearts to that of pulsars. Sometimes the signal is self evident - like in the seasons of the year, sometimes it is very faint, buried in noise and hard to detect - like, for example, that of the luni-solar and solar cycles in air temperature records [16]. Sometimes we see its pattern and are not sure of its mechanism - like in the 11-year sunspot cycle; sometimes we know the mechanism but the pattern of its imprint remains hidden - as it long was the case with climatic signals of the orbital motions of the Earth.

The last two examples are exceptionally pertinent to our topic and warrant a closer look.

Fitting the sunspot fluctuations with periodic functions has probably been the first and most notorious example of 'geophysical time-series modelling' and so were the attempts to correlate them with everything from rainfall to manufacturing and wars [17]. However, since neither their own mechanism was understood, nor any physical mechanism behind the correlations could be contemplated, none of the 'modelling' had any predictive value; its only value - as Kendall and Stuart might have agreed - was in **asking questions** to be answered by the concrete sciences. And some answers have now started to emerge - not from 'significance' tests of peaks in spectrum plots and correlation coefficients but from 'impartial

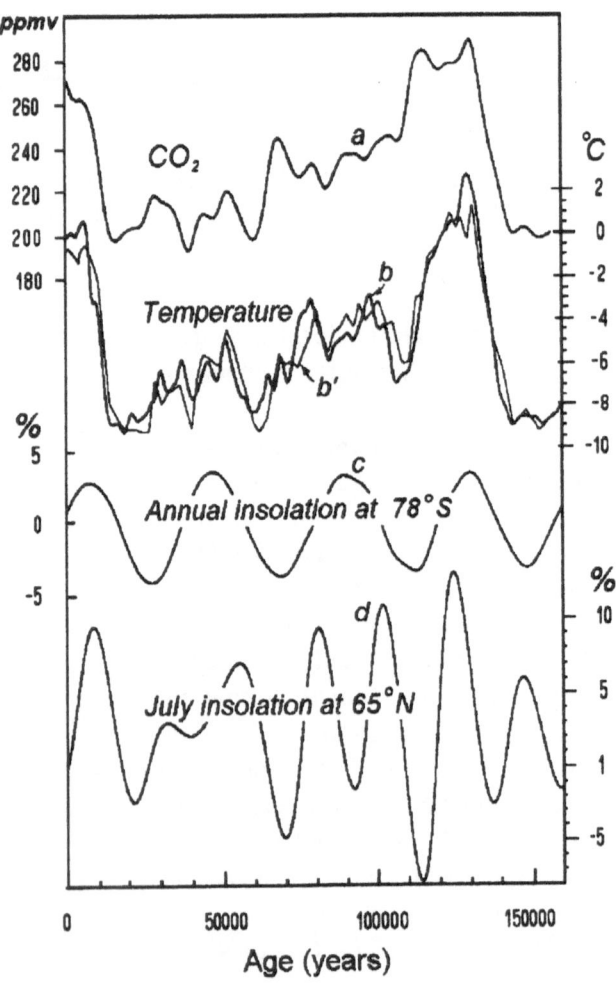

Figure 12. Vostok-related geophysical data (adapted from [20]).

investigations of the nature of things' - in this case, things like sun's angular momentum, solar wind, electrical interactions between the magnetosphere, ionosphere and higher atmosphere, etc. True, answers concerning correlations with stock market fluctuations and military flareups in the Balkans are still somewhat fuzzy, but those about connections to global temperature and sun-weather coupling show promise [18, 19].

One road to finding the signals of orbital motions of the Earth in climate fluctuations was, as is commonly known, shown by the Milankovich theory which has been widely applied in paleoclimate research. A well known example is the paleo-temperature record derived from the Vostok ice core where the direct reconstruction, b, based on the isotope content of the ice (Fig. 12, adapted from [20]) has been closely matched by the reconstruction, b', based on the content of atmospheric CO_2 of the ice, a, and orbital forcing represented by two relevant insolation records, c and d.

5.2 ACCUMULATION

To navigate in the political mine fields of the early 1950s (the physical mine fields at the edge of my home town was a different matter altogether), one had to observe certain signposts, many of which came in the form of definitions. One which I still remember and on which I shall venture an improvement today was Lenin's definition of motion as 'the form of the existence of matter'. I am hereby proposing a more general and profound definition extending far beyond matter: 'Accumulation is the form of existence of everything'. From elementary particles which accumulate to form atoms, mice and galaxies, to letters which accumulate to form words, definitions, bibles and German encyclopedias; from pennies and dollar-bills which accumulate to make the fortunes of Bill Gates, the Vatican and the IMF, to water molecules which form droplets, puddles and oceans; from small stresses which accumulate to produce headaches, divorces, massacres, wars and earthquakes, to small lies which may accumulate to impeachable offenses of American presidents.

These all are 'facts of life' rather than 'problems' requiring 'theoretical solutions'. They simply 'are' and life does 'solve' them one way or other. Problems with 'accumulation' start when man engages in 'theoretical accumulation' which he calls Integral Calculus. The point is that mathematical integration can cause a **change in the appearance** while the real process, its **substance, remains unchanged**. Such a change in appearance is illustrated in Fig.13 for two functions, the Dirac delta function and the cosine function, which are inherent to many geophysical processes (e.g. a flood, an earthquake; seasonal temperature variation). Note the conspicuous difference between the pattern of the integrand f and the integral g: in the first case, a discrete pulse (a one-time event) gives rise to a step function (appearance of an 'intervention' with lasting consequences; cf.

122

Fig.1) and, in the second, while the resulting function appears the same as the original one, its phases are shifted: the cosine has changed into a sine.

In such changes of pattern lies the basis of the 'mischief factor' mentioned in connection with eq. (1) (which of course is just a discrete equivalent of 'continuous' integrations depicted in Fig. 13): The pattern of a plot of a time- series does dot tell us how it has arisen; for example, how much it is due to a (real, physical) integration

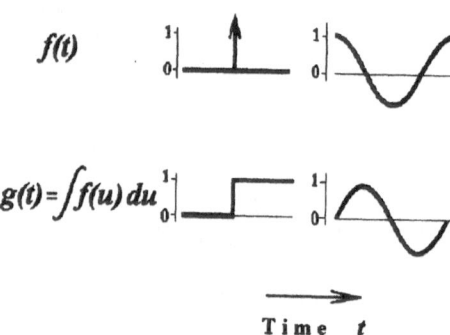

$$f(t)$$

$$g(t) = \int f(u)\, du$$

Time *t*

Figure 13. Examples of transformations of functions by integration.

done by nature and how much it reflects a ('virtual', mathematical) integration done by an analyst. When the pattern itself is used for making inferences about the real world, the difference can be enormous since in one case it reflects reality but in the other it is just an illusion.

To illustrate the point, consider the two top time series in Fig. 14 which depict 66-year records of mean annual levels in Lake Ontario and of its annual outflows which form the St. Lawrence River and are an almost-linear function of the lake level [10]. No hydrologist, climatologist or limnologist would represent lake levels by their residual mass-curves since their cumulative nature is obvious - being the product of water accumulation (=integration) in the lake, they already have a strong, and real, 'mass-curve component'. On the other hand, hydrologists and climatologists generally have no hesitation to represent the St. Lawrence flows by their mass-curve (Fig.4) since this has been 'standard practice' for river-flow representation (Figs. 2D,3,4,7).

Lake Ontario levels 1920 - 1986

Lake Ontario annual outflow (St. Lawrence River flow) 1920 - 1986

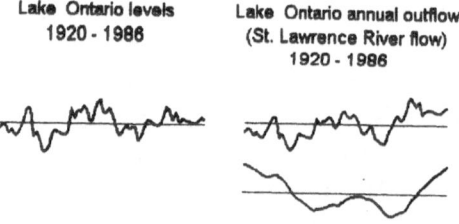

Figure 14. Historic records of annual Lake Ontario levels and outflows, and residual mass-curve of the latter [10].

However, because of their quasi-linear relationship with the lake levels, the St. Lawrence flows have almost identical pattern as do the lake levels so that they also have a strong real mass-curve component. Hence their computed, i.e. mathematically produced, residual mass-curve is already of the second order in the sense of Figs.5 and 6. Its pronounced cyclic pattern thus exaggerates the actual cyclic tendency of the flow series and its phases may be shifted with respect to those of, say, a tributary flow mass-curve. This then can encourage a false inference that, at the point of their confluence, the river and its tributary belong to different 'climatic regions'.

Cumulative mechanisms can play havoc with interpretation of geophysical records, especially when, say, climatic inferences are made from various proxy data which may contain 'mass-curve components' of orders different from the order of the process being inferred. A typical example would be the use of varves (produced by some sedimentary water-related process) as an indicator of precipitation patterns. The thickness of varves is often related to fluctuations of levels in their parent water bodies which are 'integrators' of precipitation so that the varve patterns could more likely correspond to the residual mass-curve of the precipitation process than to this process itself. A similar relationship is illustrated in Fig.15 (adapted from [21]) where the computed residual mass-curve of precipitation has been almost exactly replicated by groundwater levels actually measured in a nearby well.

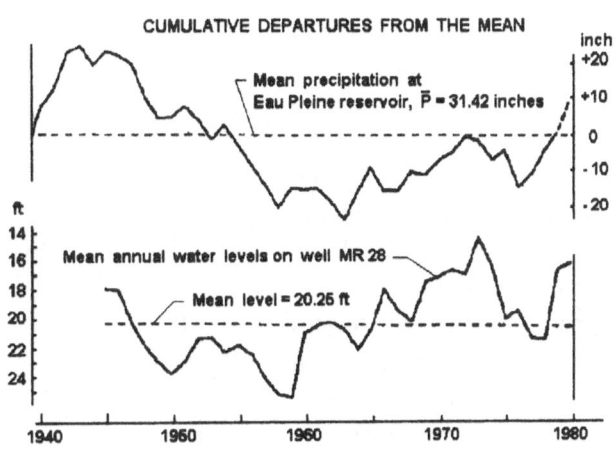

Figure 15. Groundwater levels (bottom) and residual mass-curve of precipitation at nearby station (top).

6. Inter-relationships

Acquisition of scientific knowledge follows the path *observation → description → explanation.* The last step is the most difficult one and the most misunderstood. Its misunderstanding most often arises from its confusion with description and its difficulty most often resides in nonlinearities in the explanatory relations (note the 'Kisiel prayer' quoted above) and in the multiplicity of the explanatory factors.

The substitution of description for explanation is very common in statistical analysis where the confusion is reinforced even by the standard terminology. For example, it is common practice to talk about the percentage of 'explained variance' where what is meant is its percentage which is 'accounted for'. The explanation of why this percentage is accounted for by a given 'model' is often not sought and the question is not even asked or, worse, sometimes the question is not even understood when it is asked!

124

The practice is also common in correlation analysis where a high correlation may be claimed to 'explain' a relationship. Kinsman [22] put his finger on this sore spot a long time ago: "A fallacy that is still far too prevalent in geophysical papers is the inference from a high correlation to cause and effect". This fallacy still thrives today, for example in the rather fashionable business of identifying various 'teleconnections'. Instead of the correlations being used as a legitimate way of **asking the questions**, they are often presented as **providing the answers**. Kinsman exposed this fallacy by a telling example. He analyzed a claim, based on correlation analysis, that the number of icebergs in the Atlantic, south of latitude 48° N, can be predicted from monthly temperature anomalies at Key West, Florida. He then constructed an alternative 'model' based on the correlation between the iceberg counts and the number of commas in the paper in which the above claim was published. The 'predictions' by Kinsman's 'model' are reproduced in Fig.16. Its correlation coefficient is 0.95, as compared to 0.83-to-0.92 in the original model. So much for correlations.

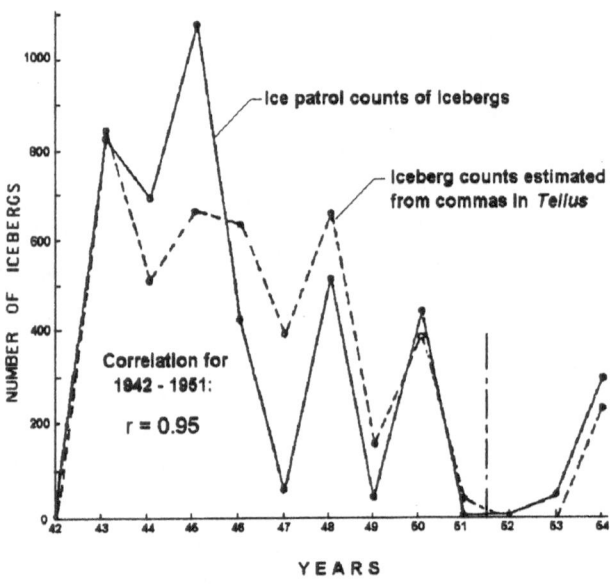

Figure 16. Annual counts of icebergs (solid line) and 'predictions' based on 'correlation model'.

One of the main reasons why relying on correlations is dangerous is the fact that physical relationships are often nonlinear. To detect and adequately describe nonlinearities is extremely difficult because, as the famous Australian statistician, the late Professor Moran, once put it [23],

> "...entirely different physical models may lead to the same stochastic behaviour and, even if they do not, the difference may be too small for discrimination in a reasonable sample ... it is clear that nonlinearity is an all pervading problem and here we are confronted, if not with a brick wall, at any rate with a hill of rapidly increasing slope".

Nonlinearities can be of vastly different forms even when the patterns of 'data plots' look almost identical. Consider, for instance, the plots in Figs.10a and 17a. They look remarkably similar and yet, their underlying relationships, though both

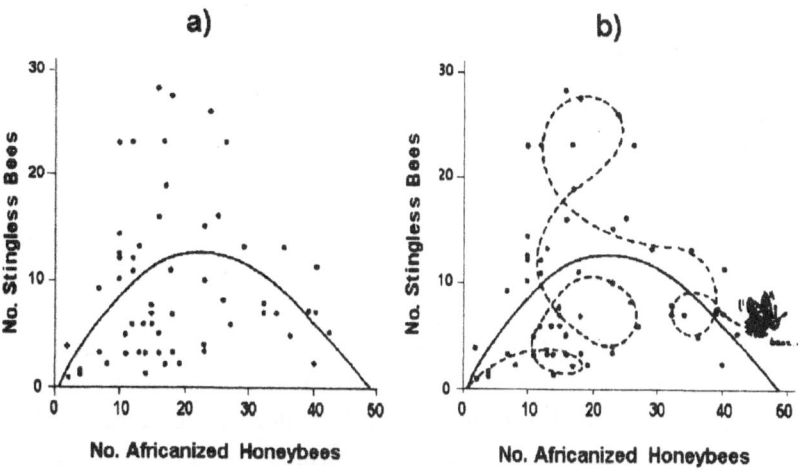

Figure 17. a) A fitted nonlinear relationship; b) an alternative interpretation proposed by a sceptic.

nonlinear, could not be more different. In the first case, much of the nonlinearity derives from transformation by storage reservoirs as illustrated in Fig.11, in the second case it is claimed to have a quadratic polynomial form obtained by 'best fit'. And while the latter may be so, in the absence of a plausible explanatory cause it certainly is legitimate to raise questions about its validity and consider alternative interpretations, though perhaps less imaginative ones than the one suggested by an inspired sceptic and reproduced in Fig. 17b [24].

In view of all the related difficulties, the reluctance of many authors to tackle causative explanations of the various inter-relationships shaping the geophysical processes is quite understandable: the multiplicity of the interacting phenomena, and the hard-to-detect and varied nonlinearities of their relationships, compound the complexity of their analysis and often seem insurmountable. Those of us who have suffered bruises trying to climb the 'brick wall' of 'nonlinear reservoirs' (e.g., [25,26,27,28]), keenly appreciate Moran's lament and Kisiel's prayer and commiserate with the colleagues who, undeterred by risk of injuries, venture onto the dangerous nonlinear slopes.

In time-series analysis, nonlinearities pose a particularly difficult problem since, as Moran [23] observed, "Nearly all work done on time series has been from the point of view of spectral analysis which is a linear theory and is not invariant under nonlinear transformations of observed values". An exemplary case of these difficulties have been the investigations of the so-called 'Hurst phenomenon' observed in many geophysical, biophysical, economical and other records and challenging analysts for almost fifty years. The most conspicuous aspect of the related investigations (and the reason for the ambiguity of their results) has been, in my view [29], the almost exclusive reliance on the standard (linear) theory of time-series analysis and an almost universal disregard for the dynamics of the possible physical processes behind this phenomenon.

It is therefore gratifying to see serious work on the dynamics behind the signals that Nature is trying to communicate to us through geophysical records. And, given the fact that the hosting institute of this international workshop has been one of the pioneers of such efforts, dating back to the work of S.N. Kritskii and M.F. Menkel on water level fluctuations of large lakes and land-locked seas, it comes as no surprise that fresh efforts in this direction are well represented in the recent collection of its works [30].

The latter work is particularly relevant to the present topic since the behaviour of large lakes reflects the interaction of many mechanisms that underlie regional and continental variations of the climate. While some of the results published [30] are tentative and some may not be confirmed by further research, the important thing is that they open up new perspectives on the possible forcings and on the complexities of their nonlinear inter-relationships: and - lo and behold! - some are actually **asking questions**: For example, in what time-frame can a body of water of the order of magnitude of the Caspian Sea switch between two apparently quasi-stable states? (Khublaryan and Naidenov, in [30], p.208). Such question, intrinsically linked to nonlinearity and threshold effects, is extremely important to the dynamics of climatic changes. It immediately conjures up broader questions such as the role of the forces responsible for the sudden ancient reversals of the Earth magnetic field or of those that had caused the sudden large-scale climate changes occurring in northern Europe until 11,000 years ago, apparently as a result of abrupt flip-overs in the ocean circulation patterns in the North Atlantic [31]. It is this type of work through which a genuine understanding of the mechanisms shaping the geophysical processes and their interactions can be reached and the patterns of their time-series properly interpreted and mathematically described.

7. Conclusion

When contemplating the history of the study of geophysical and other real-life processes on the one hand, and of the development of the mathematical theory of time-series on the other, it transpires that the latter has benefited from the former much more than it has been the other way around. This is how it should be since, at least outside the sphere of religion, substance determines form. It does not reflect negatively on the mathematical theory but shows how it is enriched by the study of nature. There is nothing demeaning in the fact that, for example, the concept of Markov chain was inspired by Markov's analysis of a Pushkin poem rather than Pushkin's poems being the product of a Markov chain model; or that Hurst's engineering investigations of the Nile flow-records inspired mathematicians to conceive the fractional Gaussian noise and not vice versa.

The important point is that enrichment of theory, and of knowledge in general, comes from honest attempts to answer questions posed by the real (and sometimes even the eternal) world, not by imposing preconceived concepts on it. While this principle seems to have been clear to Confucius twenty five centuries ago, many contemporary modellers of time-series (and other statistical aspects) of geophysical processes still have a difficulty with it; however, it now is gaining ground, however slowly, even in this area and a genuine understanding of the evolution of geophysical processes, including the climate, has started to emerge.

8. References

1. Kempthorne, O. (1971) Probability, statistics and the knowledge business, in V.P. Godambe and D.A. Sprott (eds.), *Foundations of Statistical Inference*, Holt, Rinehart & Wilson, Toronto, pp. 470-492.
2. Klemeš, V. (1978) Physically based stochastic hydrologic analysis, *Advances in Hydroscience* **11**, 285-355.
3. Klemeš, V. (1982) Empirical and Causal Models in Hydrology, in *Scientific Basis of Water Resource Management*, National Academy of Sciences, Washington, D.C., pp. 95-104.
4. Klemeš, V. (1997) Of carts and horses in hydrologic modelling (Guest editorial), *J. Hydrological Engineering* , ASCE, **2**, 2, 43-49.
5. Klemeš, V. (1999) Keeping techniques, methods, and models in perspective (Editorial), *J. Water Planning and Management*, ASCE, July/August, 151-155.
6. Hipel, K.W., Lennox, W.C., Unny, T.E., and McLeod, A.I. (1975) Intervention Analysis in Water Resources, *Water Resources Research*, **11**, 6, 855-861.
7. Todini, E. and O'Connell, P.E. (eds.) (1979) *Hydrological simulation of Lake Nasser*, Vol.1, IBM Italia Scientific Centers & Institute of Hydrology, Wallingford, UK.
8. Williams, G. R. (1961) Cyclical variations in world-wide hydrologic data. *J. Hydraulic Engineering*, ASCE, **87** (HY6), 71-88.
9. Shiklomanov, I.A. and Markova, O.L. (1987) *World Water Resources* (in Russian), Leningrad.
10. Klemeš, V. and Klemeš, I. (1988) Cycles in finite samples and cumulative processes of higher orders, *Water Resources Research*, **24**, 1, 93-104,.
11. Lalykin, N.V. (1975) Regionalization of the USSR based on co-phased annual river runoff fluctuations (in Russian), *Trudy IV Vsesoyuz. Gidrol. S'ezda*, 3, 307-316.
12. Harrington, J.B. (1987) Climatic change: a review of causes. *Can. J. Forest Research*, **17**, 11 1313-1339,.
13. Kisiel, C.C. (1967) Transformation of deterministic and stochastic processes in hydrology. *Proc. Int. Hydrol. Symp.*, Vol. 1, Fort Collins, Colo., pp. 600-607.

14. Klemeš, V. (1986) Dilettantism in hydrology: Transition or destiny? *Water Resources Research*, **22**, 9, 177S-188S .

15. Klemeš, V. (1983) Conceptualization and scale in hydrology. *J. Hydrol.*, **65**, 1-23.

16. Currie, R.G. (1983) Luni-solar 18.6- and solar cycle 10-11- year signals in USA air temperature records, *Intern. J. Climatol.*, **13**, 1, 31-50.

17. Dewey, E.R. (1971) *Cycles*, Hawthorn books, New York.

18. Labitzke, K., and van Loon, H. (1988) Associations between the 11-year solar cycle, the QBO and the atmosphere. *J. Atm. & Terr. Phys.*, **50**, 3, 197-206.

19. Bering, E.A. et al. (1998) The Global Electric Circuit, *Physics Today*, **51**, 10, 24-30.

20. Lorius, C. (1988) Antarctic ice core: CO_2 and climate change over the last climatic cycle. *Eos*, June, 681-684.

21. Zaporozec, A. (1984) Statistical evaluation of the effects of precipitation on groundwater levels, in W.H.C. Maxwell and L.R. Beard (eds.), *Frontiers in Hydrology*, Water Resources Publications, Littleton, Colorado, pp. 246-261.

22. Kinsman, B. (1957) Proper and improper use of statistics in geophysics. *Tellus*, IX, 3, 408-418,

23. Moran, P.A.P. (1975) The future of stochastic modelling, *Proc. Internat. Congress of Mathematicians,Vancouver, 1974*, Canad. Math. Congress, pp.517-521.

24. Weber R.L. (1982) *More Random Walks in Science*, The Institute of Physics, Bristol, pp. 184-185.

25. Klemeš, V, (1970) Negatively skewed distribution of runoff, *Symposium of Wellington (NZ)*, Publ. No. 96 IAHS, pp. 219-236.

26. Klemeš, V. (1982) Stochastic models of rainfall-runoff relationship, in V.P. Singh (ed.), *Statistical Analysis of Rainfall and Runoff*, Water Resources Publications, Littleton, Colorado, pp. 139-154.

27. Frolov, A.V. (1985) *Dinamiko-stokhasticheskie modeli mnogoletnikh kolebaniy urovnia protochnykh ozer*, Nauka, Moscow.

28. Unny, T.E. (1987) Solutions to nonlinear stochastic differential equations in catchment modelling, in I.B. McNail and G.J. Umphrey (eds.), *Stochastic Hydrology*, Reidel, Dordrecht, pp. 87-111.

29. Klemeš, V. (1974) The Hurst phenomenon: A puzzle? *Water Resources Research*, **10**, 4, 675-688.

30. Khublaryan, M.G. (ed.) (1994), *Vody sushi: Problemy i resheniya*. Institut Vodnykh Problem, Moscow.

31. Oldfield, F. (1998) For research on climate change, past is key to future. *Eos*, **79**, 41, 493-494.

ON THE DEVELOPMENT OF THE STOCHASTIC APPROACH IN HYDROLOGY

D.YA. RATKOVICH
Water Problems Institute, Russian Academy of Sciences,
Gubkina st. 3, Moscow, 117971 Russia

1. Introduction

River runoff, as with other hydrometeorological processes, experiences long-term and annual variations that control the hydrological regime of water bodies and their impact on the environment. Runoff variations govern aquatic ecosystems, available water resources, and determine the measures required for ensuring water supply and engineering protection against the adverse effects of water. Now, at the very end of the twentieth century it is not necessary to prove that application of statistical methods for studying natural phenomena, especially in hydrometeorology, is quite fruitful. Techniques of mathematical statistics are among the main methods in this field.

Distinguished scientists such as H.A. Einstein, N. Wiener, O. Reynolds and G. Taylor created the theoretical basis of present-day hydrometeorology at the beginning of the century. Valuable contributions were made also by the Russian scientists A.N. Kolmogorov, A.A. Friedman, L.V. Keller, E.E.Slutcki and their followers, as well as by S.N. Kritskii and M.F. Menkel. The latter two are the founders of stochastic hydrology in Russia and contributed to the solution of many applied problems.

The principal problems of modern stochastic hydrology are the following:
- To give a more precise definition of the stationarity of hydrometeorological time series and to develop procedures for revealing statistically significant trends in time series;
- To generate statistically homogeneous ensembles for probabilistic analysis, and to estimate the effect of correlation. These problems are met, for example, when classifying data on maximum river discharge for the establishment of a stochastic model and when estimating the skewness of a distribution, which in its turn has a great impact on the estimates of the maximum discharge;
- To find or to give a more precise definition of the unconditional probability distributions of all hydrometeorological characteristics including those affected by economic activities (such as regulation of river discharge by reservoirs, and also long-term regulation; the effect of water abstraction on the level of water bodies;

129

M.V. Bolgov et al. (eds.), Hydrological Models for Environmental Management, 129–137.
© 2002 *Kluwer Academic Publishers. Printed in the Netherlands.*

random variations in water consumption; and, last but not least, water withdrawals for irrigation);

▪ To determine the character (type) of serial correlation in time series and cross-correlation between different time series; the limits of possible application of simple Markov chains; and "long memory"-correlation within the series;

▪ To describe in probabilistic terms non-stationary hydrometeorological processes, including those reflecting the effect of climatic change;

▪ To develop a theory of management of water resources and engineering systems, including the utilisation of surface and ground water;

▪ To describe in probabilistic terms the transport of dissolved solids as well as bed loads and suspended loads in streams.

The environmental processes associated with hydrology should be studied with the use of stochastic analysis. It is possible that the probabilistic methods in this field will require further development.

Successful solution of the problems enumerated above as well as many other problems of stochastic hydrology requires development of new methods of probability theory and mathematical statistics and also stimulates such development. This refers, in particular, to the methods of the theory of non-Gaussian, periodically correlated (harmonic) random processes, and the problems of reliability of stochastic models and their parameters. Here, the focus will be on the problems of variations in the hydrological processes, their regularities and mathematical modelling, and the possible use of such models in hydrological calculations.

Many stochastic models have been suggested for the description of hydrometeorological processes both in Russia and in other countries. In some of these models, the variables are assumed to follow the normal distribution or a distribution similar to it, while other models require a preliminary normalization of the random variables involved. Some authors suggest transforming similar series into exponential rather than normally distributed ones. The processes including a component accounting for regular seasonal variations were subjected to processing in order to eliminate this component by means of subtracting its values averaged over relevant intervals. Smoothing of the time series was also proposed in some cases. For rivers with a mixed source of supply, threshold models may be used, which is equivalent to the use of truncated distributions.

In this paper emphasis is put on studies performed by scientists from Russia. These are less known outside Russia though they represent a well developed school in stochastic hydrology concerning both theoretical and applied aspects.

The problem of stochastic modelling of hydrometeorological processes involves a series of basic principles that are presented below.

2. Climatic stationarity

The prediction of possible variations in hydrometeorological processes is of great economic and environmental importance. On a short-term basis, only the general character of seasonal variations in hydrometeorological processes can be forecast. As to the long-term trends, they are the result of on-going climate change and are the focus of present scientific discussions. When no such trends are present, the problem reduces to the analysis of the hydrometeorological regime under stationary conditions.

In this paper, the term *climatic stationarity* has a restricted meaning. It is understood as the possibility to extend the regime regularities from the period of instrumental observations (the past 50 to 100 years) to the period for which the requirements to the regime of water objects and water resource systems can be foreseen (30 to 50 years). Thus, the term *stationarity* covers a period of up to 1.5 centuries. Analyses of a large volume of observations of runoff and precipitation showed that the number of time series with statistically significant trends is not too large and does not exceed the proportion acceptable for the relevant significance level. This allows us to accept as a first approximation the hypothesis of stationarity of the time series of annual runoff and annual precipitation. One should bear in mind that in a stationary time series of annual values, a decrease in the step (with the seasonal variations taken into account) results in the loss of stationarity, i.e. the probability distributions change within a year.

3. Theoretical distribution functions

A broad spectrum of probability distributions is used to describe the variation of hydrometeorological variables. The normal distribution adequately well describes the water level fluctuations in drainless reservoirs and the variation in evaporation and precipitation for areas where the coefficient of variation C_v does not exceed 0.1. A good description of the annual runoff can be obtained, as a rule, using the gamma distribution, for which the fixed relation $C_s = 2C_v$, where C_s is the coefficient of skewness, is known to hold. However, for some areas (e.g. Northern Kazakhstan, the Far East) this relationship turns into $C_s = 2.5\ C_v$. This makes it necessary to turn to the three-parameter modification of the gamma distribution suggested by Kritskii and Menkel [4]:

$$f(x) = ax^b e^{-cx^m},$$ (1)

where a, b, c, and m are numeric parameters that can be expressed in terms of the mean value $\bar{x}$, C_v, and C_s/C_v.

This distribution is widely used in Russia and has been recommended in numerous manuals. The three-parameter gamma distribution can be successfully used to describe variations in the minimal runoff as well as the annual amounts of precipitation and evaporation for $C_v >0.1$. It can be applied for the approximation of all water budget components, when the observation interval is less than a year (a month, ten-day or five-day period).

A large number of different types of probability distribution have been suggested for the description of river runoff. This abundance is not justified, because all these distributions yield similar results except for the zones near the extremes, which are of no practical interest for the annual runoff. The problem of the maximum water discharge is of particular importance. Maxima can have different causes, e.g., rain showers and snow melt. The time series of such annual maxima will have a change in its gradient with, as a rule, a steeper slope in the zone of the maximum values. Under such conditions it appears to be impossible to describe the probability curve by a single distribution law throughout its range. It has been found more effective to use truncated distributions with a "heavy-tail" distribution (e.g., Pareto distribution) used as an additional term.

4. Serial correlation

An important characteristic of time series is the serial correlation. This correlation is not observed only in the series of maximum annual water discharges. The serial correlation in the time series of precipitation and evaporation from water surfaces is close to the statistical significance threshold being around 0.1. A corresponding correlation for evaporation from the land surface has not been studied yet. For the annual runoff series the autocorrelation increases on the average from 0.1 in areas with water abundance to 0.5 in arid areas. Mean annual levels of inland reservoirs are an example of series with high correlation. In such series the autocorrelation increases with the shore steepness and decreases if there is an outflow of water from the reservoir. In the Caspian Sea, which is drainless, the autocorrelation coefficient r equals 0.98, and in the Ladoga Lake, with water flowing through it, r equals 0.7. Processes with high correlation between adjacent terms require the use of second-order Markov chains (in the case of reservoir water level) and special modifications of the simple Markov chain (in the case of series of hydrometeorological indices with monthly or ten-day periods) [7].

5. Non-linear correlation

The type of correlation is also of importance. Linear correlation is the best developed one most widely used in applications. However, in the case of hydrometeorological processes, it was found to disagree with observational data. It was suggested [5] to apply the linear correlation to the distribution functions rather than to the random variables themselves. The mathematical apparatus for this correlation was developed by Sarmanov [10].

Analyses of observational data for more than 400 rivers all over the world [5] showed that a distinct correlation can be seen between the coefficient of correlation r and the specific discharge rate of a river q, as can be seen from Table 1.

TABLE 1. Dependence of the autocorrelation coefficient on the specific runoff

q, $l/(s\ km^2)$	>20	20–10	10–4	4–1	<1
r	0.1	0.2	0.3	0.4	0.5

In the concept used the correlation between random variables is not linear, and the regression line tends to a certain limit. This pattern is in better agreement with physical notions.

Judging from the form of prismatic diagrams of the two-dimensional probability density, the suggested nonlinear correlation applied to the water budget components yields the results that are in a good agreement with field data. In particular, the prismatic diagrams are symmetric with respect to the diagonals, which is also typical for the observation data.

The experimental parameters of the conditional probability distributions are in a good agreement with the theoretical values for the suggested model provided that corrections are introduced to eliminate systematic errors in sample parameter estimates.

6. Patterns of variability

A typical feature of annual precipitation is a narrow variation range, a common interval of C_v variations being 0.2 to 0.3. Annual evaporation from water or from the land surface varies within a still narrower interval. Therefore, the variations of these characteristics can be described as a first approximation by the normal distribution, and in the case where a more accurate description is needed the Kritskii–Menkel distribution can be used [4]. The relationship $C_s = 1.5\ C_v$ [6] holds for precipitation whatever its magnitude and irrespective of the physiographic zone. The C_v of evaporation from a water surface is commonly less than 0.2, and C_s / C_v is about 2–3 and only the regional average of C_s/C_v can be evaluated. The skewness of the distribution of evaporation from the land surface is still to be studied.

The levels of inland water bodies, the terminal elements of river systems, are the last hydrometeorological characteristics that will be considered. Due to their high autocorrelation, even the longest observation series (100–150 years) are equivalent to only a few independent observations, which makes even an approximate determination of the type and parameters of their distribution impossible [6]. As seen from the histogram of a simulated series of the Caspian Sea levels, the distribution of this characteristic in a series with $N = 100\ 000$ terms can be adequately described by the normal law. However, in the case of a relatively short series, e.g. the available 150-year-long observation records, the histograms can have a widely varying character, some being bimodal though unimodal distributions are most common.

Some studies of the Caspian Sea are based on the results of an indirect reconstruction of the sea level series with a length of up to several hundred years. Such series cannot be used to assess the probability distribution, because climate changes within this period make the obtained series non-stationary. The way out is to use the water balance equation, which allows obtaining the required distribution and its numerical characteristics from the solution to the relevant dynamical–stochastic equation.

In the case of flow through a water body, the critical factor is the rate of increase in the water outflow from the water body when the water level rises. The greater this rate and the role it plays in the water budget, the greater the deviation of the probability distribution of water levels from the normal. For low levels the distribution is limited by the zero depth value in the outflow section. It can be approximately described by the Kritskii–Menkel distribution. Analysis of the series for throughflow reservoirs [12] yields an estimate of the C_s/C_v ratio in the range 2 to 5.

Seasonal variations were found to be the only characteristic of the hydrometeorological regime reflecting the regularities in temporal variations. The absence of seasonal variations should be regarded as an exception. For example, no seasonal runoff variations are observed in the rivers of the Caucasian coast of the Black Sea, where rain-induced peaks are spread throughout the year without any visible seasonal variation.

In the high-water periods, attenuation of the autocorrelation functions for the runoff values for ten-day periods is more rapid than that typical for the simple Markov chain. This is also valid for monthly floods characteristics with different regularities. For example, their attenuation in the Ob River is more rapid than that for power law, while in the Volga River, the attenuation is slower. Strictly speaking, the power law is inadequate in all cases; therefore, the application of the simple Markov chain requires some corrections.

A strong correlation is present only for successive months. For the lag values greater than one month, the values of $r > 0.5$ can be expected only in autumn and winter. No statistically reliable correlation can be observed in the mid-flood periods. As shown by Bolgov [7], river runoff should be regarded as a non-stationary random process with periodically changing parameters, when simulated with the interval of less than one year. The periodical correlation structure of the time series can be taken into account based on the periodic autoregressive models of different complexity (periodic autoregression models, PAR). The form of the sample autocorrelation functions for river runoff series for the former USSR territory indicates that the order of the model complexity should be selected individually for each month. In the case of significant autocorrelation between the runoff of successive years, the complexity of the models for individual months should be chosen in such a way that the coefficient of correlation between the annual values is preserved.

7. Monte-Carlo methods

The Monte Carlo method contributed largely to the introduction of probabilistic calculation techniques into engineering practice. Its application in Russian hydrology is due mainly to the works by Svanidze [11]. Two classes of problems benefit from the use of this method:
- Artificial series, with the number of terms varying from several thousands to tens of thousands, are first generated according to the adopted stochastic model. They are used then to examine the peculiarities in the series structure such as alternation and grouping of terms with specified values, ascending and descending trends, and so on.
- A number of observation series of a limited length of water budget components are used to examine the variations in the budget and to reveal water surplus or deficit, the extent of water yield failure, etc. The water budget components, as a rule, incorporate water consumption, for which the ascending trend is typical. It is worth noting that some types of water consumption are themselves of a stochastic character and require a relevant description. However, no studies of this kind are available now, and this problem is still to be investigated. Therefore, time series of water consumption are specified by their mathematical expectations.

8. Important applied problems

Among important applied problems of stochastic hydrology we would like to name specially the following:
- Evaluation of the distribution parameters. A considerable contribution to the assessment of the hydrological aspects of this problem in Russia was made by Blokhinov [1] and Rozhdestvenskii [9]. Blokhinov is known for his studies concerning the application of the maximum likelihood method for the evaluation of parameters of the Kritskii—Menkel distribution. It was proved that for $C_v < 0.5$, the method of maximum likelihood, being cumbersome, has no advantages over the common method of moments.
- The problem of modelling an ensemble of series with specified numerical parameters and coefficients of mutual correlation and autocorrelation. A simple solution was developed and applied in practice by Ratkovich (1983). The idea behind this approach was to go over from the mutually correlated to the independent time series. There exist also other, much more complicated methods, but they have no advantages compared to the one mentioned above.
- Improvement of the parameters and extension of the hydrological series using the data for similar water objects with longer observation records. The difficulty in this task is that using the regression approach, one should not ignore the random scatter typical for the series that are extended [8].

9. Conclusions

1. In recent decades important results were obtained practically in all branches of stochastic hydrology both in theory and practice. This refers to unconditional probability distributions of hydrometeorological variables and methods of evaluation of their parameters, the character and type of correlation in the time series of these variables, types and parameters of conditional probability distributions, and application of the Monte Carlo method for solving the problems of hydrology and water supply. However, in a wide range of problems the results are still tentative and require improvement.

2. An important feature that becomes even more common in stochastic hydrology is the development of models to meet specified requirements following from analysis of field observations rather than creating models with subsequent checking of their consistency with observation data.

3. The most important problems include:

- development of stochastic models for the hydrometeorological elements in cases where there is no adequate solutions or the available solutions fail to account for the peculiarities of seasonal variations;
- development of stochastic models of water consumption taking account of climatic (weather) conditions;
- development of stochastic methods for the assessment of water quality and aquatic ecosystems and problems connected to the transport of non-reactive pollutants by water both in suspended and bed load form.

10. References

1. Blokhinov, E.G. (1974) *Raspredelenie veroyatnostei velichin rechnogo stoka (Probability Distribution of River Runoff Values)*, Nauka, Moscow (in Russian).

2. Bolgov, M.V. (1997) O stokhasticheskoi modeli ekstremal'nykh raskhodov vody (On a Stochastic Model of Extreme Water Discharge Values). In: *"Analysis and Assessment of Natural Risks in Construction"*, Proc. Intern. Conf. PNIIS Gos. Kom. Stroit. Ross. Feder., 32 – 35. (In Russian.)

3. Frolov, A.V. (1985) *Dinoamiko-stokhsticheskie modeli mnogoletnikh kolebanii urovnya protochnykh ozer (Dynamic–Stochastic Models of Long-term Variations in the Level of Flow-Through Lakes)*, Nauka, Moscow (in Russian).

4. Kritskii, S.N. and Menkel, M.F. (1981) *Gidrologicheskie osnovy upravleniya rechnym stokom (Hydraulic Principles of River Runoff Management)*, Nauka, Moscow (in Russian).

5. Ratkovich, D.Ya. (1976) *Mnogoletnie kolebaniya rechnogo stoka (Long-Term Variations in River Runoff)*, Gidrometeoizdat, Leningrad (in Russian).

6. Ratkovich, D.Ya. (1993) *Gidrologicheskie osnovy vodoobespecheniya (Hydrological Principles of Water Supply)*, Water Problems Institute, Russian Academy of Sciences, Moscow (in Russian).

7. Ratkovich, D.Ya. and Bolgov, M.V. (1997) *Stokhasticheskie modeli kolebanii sostavlyayushchikh vodnogo balansa rechnogo basseina (Stochastic Models of Variations in the Components of Water Budget in a River Basin)*, Water Problems Institute, Russian Academy of Sciences, Russian Foundation for Basic Research, Moscow (in Russian).

8. Ratkovich, L.D. (1983) O vosstanovlenii korotkikh ryadov stoka po analogam (On Reconstruction of Short Runoff Series Based on Analogues), *Vodn. Resur.* 5, 26-44.

9. Rozhdestvenskii, A.V. (1977) *Otsenka tochnosti krivykh raspredeleniya geologicheskikh kharakteristik (Assessment of the Accuracy of Distribution Curves for Hydrological Characteristics)*, Gidrometeoizdat, Leningrad (in Russian).

10. Sarmanov, I.O. (1968) Postroenie korrelyatsii mezhdu ravnomerno raspredelennymi sluchainymi velichinami (Construction of Correlation between Uniformly Distributed Random Variables), *Tr. Gos. Gidrol. Inst.* **160**, 81 - 89(in Russian).

11. Svanidze, G.G. (1964) *Osnovy rascheta regulirovaniya rechnogo stoka metodom Monte-Karlo (Principles of Calculation of River Runoff Management by Monte Carlo Method)*, Metsniereba, Tbilisi.

STOCHASTIC MODELS OF THE LONG-TERM RIVER RUNOFF FLUCTUATIONS WITH APPLICATION IN WATER CONSTRUCTION AND MANAGEMENT

A.V. ROZHDESTVENSKY
State Hydrological Institute,
St. Petersburg, Y.O., 2ᵃ Line, 23, Russia

1. Introduction

The approach of using experimental statistical methods (Monte Carlo methods) is of great value in solving applied hydrological problems for cases where no analytical solution exists. The article describes procedures for model formulation and their application for estimating errors, redefining statistical tests in cases of skewed and/or dependent data and for elaborating design practices.

2. Background for adopting a stochastic model

The simple Markov chain is adopted herein as the basic stochastic model of the long-term fluctuations in annual river runoff. The main parameter of this model is the lag-1 coefficient of autocorrelation of runoff for consecutive years. The range of variation of this parameter was studied for a large sample of river runoff records covering the territory of the Russian Federation (former USSR), as part of a larger dataset for rivers of the world.

The autocorrelation coefficients for different time lags of annual runoff and their accuracy were evaluated using the following procedures:

1) estimation of the standard error in the empirically estimated autocorrelation coefficients;
2) analysis of the homogeneity of the autocorrelation coefficients across space based on long records of observations of annual river runoff;
3) analysis of the stability in the autocorrelation coefficients for the different time periods;
4) comparison of empirical autocorrelation coefficients with those obtained from statistical experiments with models having a known correlation structure.

The standard error $\sigma_R(\tau)$ of the lag-τ correlation coefficient $R(\tau)$ in case of normally distributed data and a record of n years is:

$$\sigma_R(\tau) = \frac{1 - [R(\tau)]^2}{\sqrt{n - \tau - 1}},$$

(1)

where n is the size of time series and τ is the lag in years.

Statistical experimental tests have shown that the accuracy when using the above

139

M.V. Bolgov et al. (eds.), Hydrological Models for Environmental Management, 139–145.

equation also for moderately skewed data ($Cs \leq 2Cv$; $Cv \leq 1$) and for correlated data (R $(1) \leq 0.5$) for $n \leq 50$ is satisfactory. Some restrictions obtained for the application of eq. (1) are not of great importance for the analysis of random errors in estimates of autocorrelation coefficients.

Numerous computations applying eq. (1) to the longest series of annual river runoff for $\tau > 1$, has shown that the standard errors of the autocorrelation coefficients are the same order of magnitude as the absolute values of the autocorrelation coefficients themselves. We can thus conclude that the variation in empirical values of autocorrelation coefficients is dominated by random fluctuations in the selected data and do not reflect the pattern of long-term fluctuations in the river runoff as a whole.

The analysis of the empirical autocorrelation functions in accordance with step 2 above was carried out for 72 rivers with long-term (not less than 50 years) observation records of the annual river runoff. The results of the computations are shown in Table 1. These results prove that only lag-1 autocorrelation coefficients of river runoff, which are equal or larger 0.3, are statistically significant. For $\tau \geq 2$ the average value of the autocorrelation function goes quickly towards zero according to the formula of the simple Markov chain:

$$R(\tau) = [R(1)]^{\tau} \quad . \tag{2}$$

Empirical values of $\sigma^*_R(r)$ were calculated using the following equation:

$$\sigma^*_R(\tau) = \sqrt{\frac{\sum_{i=1}^{k}\left(R_i(\tau) - \overline{R(\tau)}\right)^2}{k}} \tag{3}$$

where k is the number of records (k=72). The corresponding theoretical values were calculated using eq. (1). It is seen in Table 1 that the individual values for the selected time lags are practically identical, which justifies an assumption of statistical homogeneity of these values.

TABLE 1. Autocorrelation functions for k=72 long-term annual runoff series (with at least 50 years of records)

Time lag (τ) years	Average autocorrelation coefficients $R(\tau)$	$R(\tau)$ according to the equation $R(\tau)=R(1)^{\tau}$	Standard error $\sigma_R(\tau)$(eq.1) for n=72*65=4680	Empirical standard error $\sigma_R^*(\tau)$ (eq.3)	Standard error $\sigma_R(\tau)$ (eq.1) for individual $R(\tau)$
1	0.30	0.30	0.01	0.11	0.12
2	0.10	0.09	0.01	0.14	0.13
3	0.00	0.03	0.01	0.13	0.13
4	-0.03	0.01	0.01	0.14	0.13
5	0.00	0.01	0.01	0.12	0.13

The assessment of the stability of the empirical autocorrelation coefficients (the third step in the analysis) is carried out by means of comparing the results of computations based on different record lengths (n=36, 60, 120) using the longest records of the rivers of the world. It is interesting to note that the variance of the empirical autocorrelation coefficients decreases regularly when the amount of data considered decreases, becoming totally similar to the ordinary calculation errors in the autocorrelation functions. So we can assume that if the amount of selected data constantly increases, the variance of the empirical autocorrelation functions decreases, which is in agreement with the two first procedures in the analysis.

The fourth step of the analysis consists in computing the autocorrelation coefficients for different time lags from synthetically generated data with known distribution and autocorrelation structure. The results of the computations performed for the modelled annual sequences of different length and for observed time series have shown that in both cases a decrease of the autocorrelation functions' ordinate with an increase of the amount of the initial data is remarkable. Numerous computations prove that for the long-term fluctuation of river runoff only the lag-1 autocorrelation of annual runoff is a significant characteristic and is caused by the annual carry over of stored water from one year to another.

Dependencies of the lag-1 autocorrelation coefficient of the runoff on different physiographic and morphometric characteristics and parameters of the distribution, which influence the volume of transition moisture in a river basin, have been established for the territory of Russia. The results are presented in Table 2. It shows the average values of autocorrelation coefficients calculated for the number of series in a respective class. However, errors in the individual autocorrelation coefficient values are very large. The statistical analysis of the initial data (more than 5000 gauges) allowed eliminating random errors in autocorrelation coefficients, or, at least, reducing them significantly. After this operation it was possible to reveal the dependencies of autocorrelation coefficient on the lake, swamp and forest percentage of a basin and also on the mean and coefficient of variation of the annual river runoff. Accepted arguments affect directly or indirectly the annual transition of water resources in the river basin.

3. Application of experimental statistics for applied hydrological problems

The approach of using experimental statistical methods (Monte Carlo methods) is very popular for solving hydrological problems in cases where no analytical solution exists. These problems include a variety of topics such as: the assessment of the accuracy of the selected statistical parameters including quantiles of a distribution of the hydrological characteristics [2]; generalization of statistical characteristics of the hydrological data in case they are correlated in time and space [1, 4]; river runoff regulation etc. The method followed is based on mathematical apparatus developed for normally distributed data containing autocorrelation, which are subsequently

TABLE 2. Dependence of the autocorrelation coefficient on physiographic and morphometric characteristics and parameters of the distribution.

Dependence on the percentage of lakes

Lake %	0-1	2	3-5	6-8	9-11	12-14	15-25
Number of series	136	19	13	5	2	3	4
$r(1)$	0.16	0.40	0.48	0.08	0.54	0.46	0.69

Dependence on the percentage of swamps

Swamp %	3-5	6-10	11-20	21-30	31-40	41-50	51-100
Number of series	454	356	316	113	63	26	37
$r(1)$	0.21	0.21	0.23	0.29	0.39	0.44	0.52
$\sigma_{r(1)}$	0.32	0.32	0.34	0.33	0.31	0.32	0.43
$Cs_{,r(1)}$	-0.40	0.00	-0.26	-0.32	0.23	-0.01	-0.53

Dependence on the percentage of forest

Forest %	0-10	11-20	21-40	41-60	61-80	81-100
Number of series	500	341	500	462	500	500
$r(1)$	0.15	0.17	0.21	0.23	0.19	0.15
$\sigma_{r(1)}$	0.36	0.34	0.31	0.34	0.34	0.32
$Cs_{,r(1)}$	0.14	-0.16	-0.09	-0.20	-0.20	0.08

Dependence on the average long-term annual runoff

Average long-term annual runoff (l/s km^2)	0-2	2-4	4-6	4-8	8-10	15-40
Number of series	476	500	500	500	500	500
$r(1)$	0.23	0.23	0.21	0.21	0.20	0.14
$\sigma_{r(1)}$	0.37	0.38	0.35	0.32	0.30	0.33
$Cs_{,r(1)}$	-0.18	-0.09	-0.14	-0.48	-0.21	-0.06

Dependence on the coefficient of variation of annual runoff, C_v

C_v	0.0-0.15	0.15	0.20-0.30	0.30-0.40	0.40-0.50	0.5-0.60	0.60-1.5
Number of series	195	500	500	500	451	212	372
$r(1)$	0.01	0.11	0.18	0.19	0.25	0.22	0.26
$\sigma_{r(1)}$	0.33	0.35	0.32	0.28	0.36	0.40	0.37
$Cs_{,r(1)}$	-0.10	-0.03	0.06	-0.36	-0.19	-0.15	-0.20

transformed into asymmetrically distributed river runoff time series. Dependencies of the autocorrelation coefficients on λ-distributed and normally distributed quantities for the Pearson type III distribution and Kritsky-Menkel three-parameters λ-distribution were obtained.

4. Random and systematic errors in the selected parameters and quantiles of a distribution

Based on a stochastic model of the simple Markov chain type and applying experimental statistics, random and systematic errors of selected parameters and quantiles of a distribution were obtained for all the ranges of hydrological parameters of the distributions. This was done for the selected data for the Pearson type III and Kritsky-Menkel distributions and for different lag-1 autocorrelation coefficients of runoff.

The first step in this procedure includes generation of a set of random numbers of length N using a known analytical distribution (for example, Pearson type III) with the parameters m_x (mean value), Cv (coefficient of variation), Cs (coefficient of skewness) and the lag-1 autocorrelation coefficient $r(1)$, which is then split into smaller samples of length n. In a second step the selected parameters m_x, Cv, Cs, $r(1)$ for each sample of length n are computed. The number of these series is l=N/n. For each sample 27 ordinates of the distribution curve are computed at a third step. Using the results obtained at the second and third steps the parameters of the distribution or the distribution functions of random errors in the selected parameters (the 4th step) and quantiles of the distribution (the 5th step) are determined.

The numerous computation results are generalized in [2]. A more flexible and physically founded system for hydrological design than the one used at present in Russia and other countries is suggested in the same paper. This system defines not only the probability value of failure, which is related to the life-time of construction, but also computes the design extreme values, which depends on the importance of the construction. To be able to achieve the required design values, random errors of the distribution quantiles need to be determined. The suggested system for hydrological design can be used for the calculation of any hydrological characteristic. For example, for the computation of the maximum discharge the upper extreme value of the calculated quantile is used, and for minimum annual and seasonal runoff the lower extreme value of calculated quantile can be used.

5. Generalization of the statistical homogeneity criteria for the asymmetrical hydrological data correlated in time and space

Assessment of the homogeneity and stability of long-term fluctuations in river runoff becomes more important due to the intensification of economic activity on rivers and in river basins. However, the available statistical homogeneity criteria are elaborated for random independent values. Using these criteria in hydrological computations may lead to wrong conclusions because of the correlation of the hydrological processes in time and space and the asymmetry of their distributions. Therefore, in "Recommendations for the statistical methods of analysis of spatial and temporal runoff fluctuation homogeneity" [1] a generalization of the parametric criteria for homogeneity was made for mean values (the Student criteria) and variances (the Fisher criteria). This was done also for non-parametric criteria, such as the Dixon and Smirnov-Grubbs for the Pearson type III distribution and the simple Markov chain, and the Kolmogorov-Smirnov criteria for the lag-1 correlation.

This task was solved using the method of experimental statistics for which, at the second step, computations of homogeneity statistics (e.g. Fisher statistics) were carried out. With this as a base, statistical distribution functions for autocorrelated runoff and with asymmetry of the initial distribution were determined. Using this latter distribution function, critical test statistics for the generalized criteria of homogeneity or goodness-of-fit can be obtained. The results of the extensive calculations are generalized in tables and diagrams given in "Recommendations ..." for the so generalised Student's, Fisher's, Dixon's, Smirnov-Grubbs' and Kolmogorov-Smirnov's statistics.

6. Development of a new system of standard routines for determination of hydrological characteristics in engineering problems

The State Hydrological Institute (SHI) started developing a new system of standard routines for the determination of design practices of hydrological characteristics for engineering problems. According to the Design standards and regulations for the Russian Federation (CNR) 10-01-94, preparation of the following Federal normative documents is foreseen:
– Design standards and regulations for the Russian Federation;
– Code of regulations for projecting and construction (CR) and normative documents for the Russian Federation;
– Regional design standards (TCS).
At present a project in the frame of CNR "Calculation of hydrological characteristics. Main regulations" is ongoing at SHI with the assistance of the Russian State Hydrometeorological University, Institute of Water Problems, Soyuzvodopoekt and Gidroproekt Institute. The regulations developed will be used for different tasks connected to land-use and projecting and exploitation of different constructions. The stochastic model of the simple Markov chain is adopted in this document for the description of long-term fluctuations of river runoff.

These standards and regulations can be applied for assuring the functioning and stable development of such branches of the economy as: pipelines for gas and oil; hydraulic constructions in rivers; nuclear and thermo-electric power plants; rail- and motorways, reclamation systems, water supply systems, planning of settlements; general planning of industrial and economic enterprises etc. They are also needed for elaboration of flood protection measures.

The novelty and scientific value of the new CNR comprise:
1. New computational methods for engineering hydrology contributing to more accurate and easier determination of hydrological characteristics for the cases when hydrological observations are available, scanty or absent;
2. A link to new research and development;
3. Involvement of the organizations, enterprises and specialists and independent development;
4. Emphasizing the role of hydrological investigations when the existing observation net is insufficient and decreasing;
5. More complete use of the existing hydrological data;

6. Reduction of the obligatory and advisory construction norms (Code of regulations and Territorial Construction Standards of the Russian Federation);
7. New approaches to engineering hydrological calculations for economic activities in river channels and river basins.

The code of regulations for projecting and construction includes technical recommendations and computational methods for assistance to follow the obligatory regulations of CNR.

The Regional Construction Standards of the Russian Federation (TSC) represent the main normative document for the projecting and construction in the territory of the Russian Federation. These standards are in agreement with CNR and they include all the necessary generalized hydrometeorological data for the Russian Federation (multiple maps, characteristics and parameters in computational schemes and formulas, numerous tables and nomograms, required in engineering hydrology).

A computerization of hydrological calculations using an interactive dialog system for PCs; an extensive use of GIS, including the construction and updating of digital maps for hydrological characteristics and runoff parameters are stipulated when preparing the Federal Code of Regulations and Regional Construction Norms for Russian Republics, Territories and Regions.

The described complex of related regulative documents is aimed at improving engineering practices and planning. Such a complex of normative documents is elaborated for the first time in Russia and will facilitate a stable development of the Russian economy.

References

1. *Recommendations for the statistical methods of analysis of spatial and temporal runoff fluctuation homogeneity* (1984), Gidrometeoizdat, Leningrad (in Russian).
2. Rojdestvensky, A.V. (1977) *Assessment of the assurance of the hydrological characteristics distribution curves*, Gidrometeoizdat, Leningrad (in Russian).
3. Rojdestvensky, A.V. (ed.) (1988) *Spatial and temporal fluctuations of the runoff of USSR rivers*, Gidrometeoizdat, Leningrad (in Russian).
4. Rojdestvensky, A.V., Ejov, A.V.and Saharuk ,A.V. (1990) *Assessment of hydrological computations accuracy*. Gidrometeoizdat, St. Petersburg (in Russian)
5. Rojdestvensky, A.V. and Tchebotarev, A.I. (1974) *Statistical methods in hydrology*, Gidrometeoizdat, Leningrad (in Russian).

STATISTICAL MODEL OF RIVER-BED STREAM: PARAMETERS OF TURBULENCE, ENERGY CHARACTERISTICS

V.K. DEBOLSKY and E.N. DOLGOPOLOVA

Laboratory of River-Bed Streams,
Water Problems Institute, Russian Academy of Science
117735 Moscow, ul. Gubkina , 3

1. Introduction

The solution of all problems of sediment transport in rivers demands a definition of the structure of the flow, which includes the characteristics of mean and turbulent motion, pressure, temperature, concentration of suspended particles and resistance of the flow. The parameters of turbulent transfer, as well as the energy dissipation, which determines hydraulic resistance of the stream, are based on the knowledge of the flow characteristics. The theory of turbulent flows includes the unclosed system of differential equations. To close this system the gradient form of turbulent diffusion is often used. This approach was developed in papers by V.M. Makkaveev [9].
The investigation of production and dissipation of turbulent energy is very helpful for description of the processes of diffusion and sediment transport. To obtain turbulent diffusion it is necessary to know the gradients of mean velocity and statistical moments of the fluctuations of the flow velocity. Production and dissipation of turbulent energy in rivers are discussed below.

2. Production of Turbulent Energy

To estimate the magnitude of production of turbulent energy T we use the following expression:

$$T = -\overline{u' v'} \frac{du}{dy},$$ (1)

where u, v are the time averaged longitudinal and vertical components of flow velocity; $\overline{u' v'}$ is the correlation between fluctuations of these components; and y is the vertical direction from the bottom.

The distribution of shear stress with depth is:

$$-\overline{u' v'} = u_*^2 (1 - y/h) ,$$ (2)

147

M.V. Bolgov et al. (eds.), Hydrological Models for Environmental Management, 147–155.
© 2002 Kluwer Academic Publishers. Printed in the Netherlands.

where u_* is the friction velocity and h is the depth of the flow.

To find the derivative of the mean velocity of the flow and correlative moments of velocity fluctuations it is necessary to know the distribution of these characteristics with depth.

The mean velocity profile can be described in two ways:

$$u = \frac{1}{\kappa} u_* \ln \frac{y}{h} + C \tag{3}$$

or

$$u = (1+n) < u > \left(\frac{y}{h} \right)^n, \tag{4}$$

where κ is the Karman's constant, C is a constant, n is the power coefficient, $<u>$ is the depth averaged velocity.

The depth distributions of flow velocity obtained by measurements can be described adequately by expressions (3) and (4) [1]. In general, one can improve the fit for the measured velocity profile by a proper choice of parameters κ and n. Based on a wide range of laboratory and field experiments we suggest the following expressions for κ and n:

$$\kappa = \frac{1}{2} \left[\left(\frac{\rho_0}{\rho_w} \right)^{1/3} \frac{d}{h} \frac{\bar{u}}{(g\nu)^{1/3}} \right]^{0.04}, \tag{5}$$

$$n = \frac{1}{5} \left[\left(\frac{\rho_0}{\rho_w} \right)^{1/3} \frac{d}{h} \frac{\bar{u}}{(g\nu)^{1/3}} \right]^{0.06}, \tag{6}$$

where $\rho_0 = \rho_s - \rho_w$, ρ_s, ρ_w are the densities of the sediment mixture and water respectively, d is the size of the sediments, ν is kinematic viscosity, and g is the gravity constant.

The fluctuations of velocity components can be described in terms of their probability distribution functions. The random character of advection of the velocity fluctuations is due to the random character of the sediment composition, local changes of the water surface level, etc.

There is a wide range of works suggesting different distributions for the fluctuations of the flow velocity [6]. As indicated in many investigations, these fluctuations can be described by different density functions, depending on the variety of characteristics of the stream, for example, bottom roughness or concentration of the sediment mixture. They vary with the size of particles of the mixture, different concentrations or flow velocities [7]. Thus we may conclude, that it is at present impossible to obtain the general density functions for the distributions of fluctuations of flow velocity for natural streams.

Experiments and their theoretical generalization play, thus, a key role in solving this problem. A generalization of experimental results is very important because for natural streams the main characteristics of the flow (depth, velocity, sediment size, slope of free surface and bottom) vary, as a rule, within a range of one order of magnitude. At the same time the accuracy of measurements of river-bed flow parameters is about 10%, and for sediment transport it reaches 50%.

Let the velocity profile of a natural stream is described by the power law, which fits adequately for rivers [11]. Then, the derivative of the mean velocity is:

$$\frac{du}{dy} = (1+n)n <u> h^{-n} y^{n-1}. \tag{7}$$

It is well known that in the main part of the flow, excluding the boundary layers, the distribution of velocity can be described both by power and logarithmic laws. Using (3) and (4), the friction velocity u_* [2] is obtained as:

$$u_* = \kappa n <u>. \tag{8}$$

The production of turbulence energy T is then obtained as:

$$T = \frac{\kappa^2 n^3 (1+n) <u>^3}{h} (y/h)^{n-1}(1-y/h). \tag{9}$$

Normalization of (9) by $h/<u>^3$ gives for the dimensionless turbulent energy production T':

$$T' = \kappa^2 n^3 (1+n)(1-y/h)(y/h)^{n-1}. \tag{10}$$

150

Eq. (10) contains only the depth of flow and the power coefficient, which is connected to the hydraulic friction of the flow. The depth distributions of the turbulent energy production for the rivers Kirzhach and Polomet are shown in Figures 1 and 2. The main part of the turbulent energy production has its origin at the bottom layer. This fact confirms the conclusion made by Townsend that "90% of energy of turbulence is produced at the layer, where the shear stress changes only by 10% " [12].

3. Dissipation of Turbulent Energy

The magnitude of energy dissipation S can be found by measuring the spectra of turbulent fluctuations of stream velocity of the river. The energy spectral density of the unit of fluid mass within the range of wave numbers dk obeys the law of "-5/3" [8], and for the energy spectral density as a function of the frequency we have:

$$E(\varpi) \sim (\bar{u}S)^{2/3} \varpi^{-5/3}. \tag{11}$$

Based on the data of frequency spectra of velocity fluctuations for the rivers Kirzhach and Polomet, the distributions of normalized energy dissipation S' through the depth, shown in Figures 1 and 2, were calculated. Normalization is the same as for T' in (5).

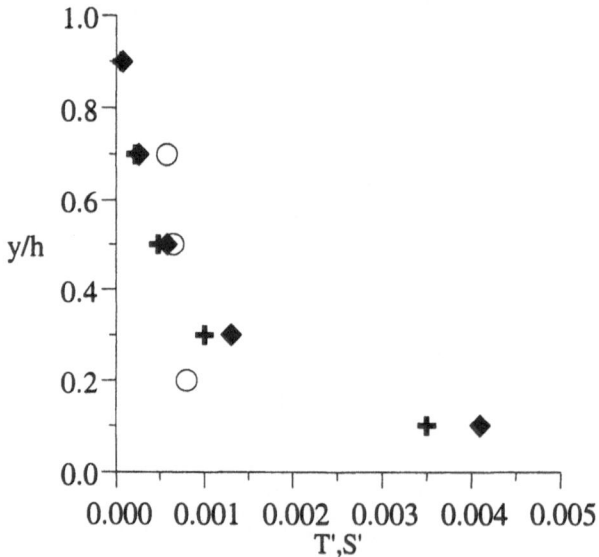

Figure 1. Distribution of turbulence and dissipation
energy through the depth of the rivers.

◆　　　T', the Kirhzach River

○　　　S', the Kirhzach River

✛　　　T', the Volga River

The integral value of T' with depth can be calculated with the help of $< T' >= \kappa^2 n^2$. The estimation of the depth-averaged turbulent energy production gives for the Kirzhach River $<T'>=0.0037$ and for the Polomet River $<T'>=0.014$. This quantity was also estimated for the Volga River with the help of the distribution of turbulent energy production with depth shown in Figure 2. The calculation for the Volga River gives $<T'>=0.0033$. Thus, the magnitude of $<T'>$ can be considered as a degree of turbidity of the flow independent on the scale of the flow. Visually, the curves in Figures 1 and 2 are similar: at the bottom $T'>S'$, at the $y/h\sim0.5$ $T'\sim S'$, and at the depth $0.5<y/h<1$ $S'>T'$.

The results of the measurements of the turbulent energy production and dissipation in the Polomet River were compared with the measurement data from the laboratory flume for the flow behind an obstacle [4]. The height of the obstacle was half of the depth of flow, and the height of the sand waves in the Polomet River was also comparable to the depth of the river ($\sim$ 1/3h). Figure 2 shows that the curves of energy dissipation for these two cases are very close. The obtained depth averaged value of energy dissipation for the Polomet River was 0.006, and for the Kirzhach River - 0.0006, i.e. the discrepancy between these values is about one order of magnitude.

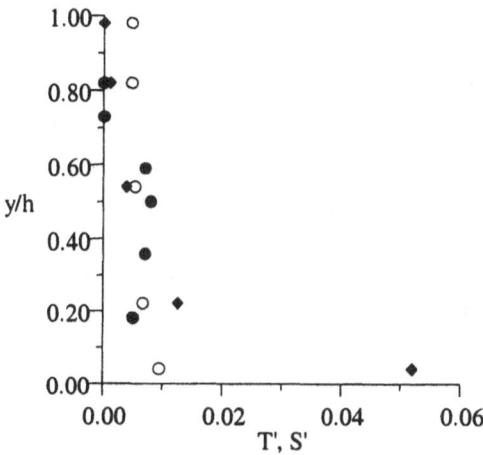

Figure 2. Distribution of turbulence
production and dissipation with depth of the flow.

◆   T', the Polomet River

○   S', the Polomet River

●   S', the flume

4. Mean Kinetic Energy

Consider evaluating the average energy characteristics of water flows. The ratio of kinetic E_k and potential E_p energy of the flow can be described by the expression:

$$\frac{E_k}{E_p} = \frac{<u>^2}{2gh}.$$
(12)

Equation (12) is a half of the Froude number. The Froude number for plain rivers is in the order of 0.1 and it can be higher than one only for mountain rivers.

The estimate of the ratio of viscous dissipation energy E_v and the total energy of the flow E can be expressed as:

$$\frac{E_v}{E} = \Delta\sqrt{\lambda},$$
(13)

where λ is the coefficient of hydraulic resistance, and the dimensionless coefficient Δ is about 5.

The expression in eq. (13) shows that the ratio of the viscous and the total energy of the flow depends on the Reynolds number in the same way as the coefficient of hydraulic resistance does. The dependence λ(Re) for the flumes is presented in [11] and for the rivers it is discussed in [3].

Let us write down the Reynolds number as:

$$\text{Re} = \frac{<u>^3}{h\left(\dfrac{<u>^2\, v}{h^2}\right)}.$$
(14)

This expression makes possible an evaluation of the ratio of the sum of kinetic energy and the energy of viscous dissipation of the flow. The energy of viscous dissipation for the natural flows constitutes about 1% of the total kinetic energy of the flow.

5. Necessary Stream Energy to Suspend Sediment

Consider now the amount of energy, which the flow expends on transportation of sediments. An expression for the carrying capacity of the flow, S_f , suggested by M.A Velikanov and later by K.I. Rossinskiy [see, e.g. 10], can be used such that:

$$S_f = 0.024 \frac{V^3}{ghw},$$

(15)

where w is the vertical velocity, and V is the velocity defined from the discharge of the stream.

From a physical point of view, this relationship is the ratio of the total energy of the flow $\left(V^3/h\right)$ and the energy, which the flow spends to keep sediments in suspension ($g \cdot w$). The concentration of the suspended sediments in natural streams is about 1%. This means that a sluggish river usually spends not more than 5% of its total energy to transport suspended sediments.

To estimate the influence of sediment transport on the change of the flow energy, let us express the shear stress in eq. (1) as a function of the turbulent viscosity V_t:

$$\overline{u'v'} = V_t \frac{du}{dy}.$$

(16)

Then, eq. (1) can be reduced to the form:

$$T = V_t \left(\frac{du}{dy}\right)^2.$$

(17)

At the same time V_t can be expressed as [5]:

$$V_t = \sigma_u^2 \theta_L,$$

(18)

where θ_L is the Lagrangian time scale of turbulence, σ_u^2 is the variance of fluctuations of stream velocity. θ_L , as a function of the Eulerian time scale θ_E , can be obtained from the following expression [4]:

$$\theta_L = \frac{\theta_E}{1.5 n \kappa}, \tag{19}$$

where n and κ are the parameters of the power and logarithmic laws for flow velocity profiles, respectively. Taking into account the relation between the Eulerian time and space scales:

$$L_E = u \theta_E, \tag{20}$$

for the Lagrangian time scale of turbulence the following relationship is obtained:

$$\theta_L = \frac{L_E}{1.5 n \kappa u}. \tag{21}$$

It has been shown in many papers ([13] among them) that the suspended sediments begin to influence the value of σ_u^2, compared to the flow free of suspension, starting with the sediment concentration of about 1%. Thus, for the majority of natural streams it is usually assumed that the variance of fluctuations of the stream velocity does not depend on the variations of the concentration of the suspended sediment. However, the scale of turbulence is a different matter. For the flow transporting suspended particles the linear Eulerian scale essentially depends not only on the concentration of the suspension, but also on the size of suspended particles. The carrier frequency of the velocity fluctuations in the main water body of natural streams is about 10 Hz [10, 13], which is about 1 cm in accord with the linear Eulerian scale. When suspension concentration is close to 1%, the linear size of the suspended particles, uniformly distributed in a volume of 1 cm^3, occupy the 0.01 part of this volume that is 0.2 cm. This is the projection of all these particles on each side of the cube with the edge of 1 cm.

If the particles are fine, they suppress turbulence in the stream, since their amount is large and the spacing between them is small. In this case the scale of the turbulence decreases in magnitude till about 0.2 cm. If the particles are big enough, then at the same concentration there will be fewer particles in the same volume, and the space between them becomes bigger. Thus, the linear scale of turbulence either appears to be close to that of the free flow, or may even increase due to the transfer of the force pulse from turbulent fluctuations to the next elementary volume. For example, for particles with diameter ~ 0.1 mm the space between them in the elementary volume is about 0.35 mm, and for particles with diameter ~ 1 mm this space is ~ 2.8 mm. In the former case 20 particles fit the linear size, while only two of them do so in the latter.

In the near bottom area of the stream the carrier frequency of the velocity fluctuations is about 1Hz [10, 13] and the velocity of the stream is much smaller. Thus, the local Eulerian scale of turbulence varies over less than one order of magnitude. On the other hand, the concentration of the suspension and the size of the particles increase in the bottom layer in comparison with the main water body of the stream. Consequently, it is impossible to make a unique conclusion about the magnitude of the change of the linear turbulence scale. However, some estimates of this change for natural streams indicate an approximate value of ~ 10%. The energy of the stream will change by the same order.

6. Acknowledgement

This work was partly supported by the RFBR (Grant No. 97-05-64592).

7. References

1. Barenblatt, G. (1978) *Similarity, Self-similarity, and Intermediate Asymptotics.* Hydrometeoizdat, Leningrad.
2. Debolsky, V.K., Dolgopolova, E.N., Orlov, A.S. (1989) Statistic characteristics of dynamics of river flows. *Hydrophysics Processes in Rivers, Reservoirs and Seas,* Science, Moscow, 50-66.
3. Dolgopolova, E.N. (1993) The dependence of Darcy-Weisbach resistance coefficient on the shape of river bed, in *Proc. of International Symp. "Runoff and Sediment Yield Modelling",* Warsaw, Poland, 285-291.
4. Fidman, B.A. (1991) *Turbulence in Nature flows,* Hydrometeoizdat, Leningrad.
5. Frost, W., Moulden, T.H. (1977) *Handbook of Turbulence Fundamentals and Applications, Vol.1,* Plenum Press, New York, London.
6. Grinvald, D.I. and Nikora, V.I. (1988) *River Turbulence,* Hydrometeoizdat, Leningrad.
7. Kril',S.I. (1985) *Two Conceptions of Theory for Sediment-suspending Flows,* Science, Moscow.
8. Landau, L.D., Lifshits, E.M. (1986) *Hydrodynamics,* Science, Moscow.
9. Makkaveev, V.M. (1971) *River Flow and Channel Processes,* Science, Moscow.
10. Mikhailova, N.A. (1962) *Transport of Sediment by Turbulent Water Flows,* Hydrometeoizdat, Leningrad.
11. Schlichting, H. (1969) *Theory of The Boundary Layer,* Science, Moscow.
12. Townsend, A.A. (1956) *The Structure of Turbulent Shear Flow,* Cambridge University Press.
13. Wang, Z., Larsen, P. (1994) Turbulence characteristics of suspended sediment flow, *J. Hydraulic Engineering, ASCE,* **120,** No. 5, 577-585.

MANAGEMENT OF WATER RESOURCES SYSTEMS UNDER NON-STATIONARY CONDITIONS

I.L. KHRANOVICH and A.L. VELIKANOV
Water Problems Institute of the Russian Academy
of Sciences: 3 Gubkina str., 117333, Moscow, Russia

1. Introduction

A water resources system (WRS) is a complex system with a lot of parameters and relationships determining its operation and development, and having complex interrelationships between its subsystems and between the system itself and other natural and economic systems. All the WRS parameters vary with the time. WRS processes and external impacts affecting it are uncertain and stochastic. Therefore, methods of WRS management are based on the general principles of managing complex systems, taking into account the characteristics of a particular WRS and adopting some simplifying assumptions.

In this paper a WRS is regarded as an ensemble of interacting elements *viz.* water sources, water users, means of water supply and management, water resources quality and regime. By water sources in the WRS is understood river runoff in one or several basins. Water users are different industrial, agricultural, municipal, power production and fishery enterprises, as well as recreation and flood control facilities. The means of water supply and management include river reaches, canals, pipelines, pumping-storage stations, reservoirs, water intake and return facilities, etc. Water quality is controlled either by water users in a process of water resources utilization, or by means of creating special water protection zones.

The main characteristic feature of a WRS is the stochasticity of the resource utilized, i.e. water in the form of river runoff. Water demands, especially in agriculture, are also subject to stochastic fluctuations. This creates certain difficulties in WRS planning such as the necessity in characterizing these fluctuations and deterministic decision-making on the regime and parameters under stochastic conditions. One of the main simplifications allowing these difficulties to be overcome is an assumption of stationarity in the WRS operation and development. This assumption is valid for a WRS whose connections with natural and socio-economic processes are weak. Until recently, practically all existing WRSs fell under this definition. Intensified economic activity led to the necessity to analyse WRSs under non-stationary conditions.

M.V. Bolgov et al. (eds.), Hydrological Models for Environmental Management, 157–168.

At present, the strategy for the rational use of water resources has to be changed due to the fact that the human impact on natural processes is now comparable to the dynamics of the latter. This is evident from the modification of water quality and quantity both under the effect of a direct utilization of water and as a result of economic activity and changes in the priorities and water user's demands for water quantity and quality. During intensified development of the national economy, when water demand exceeds the production of water resources, restructuring and reconstruction of the national economy brings with it the necessity to study WRS operation and development under non-stationary conditions. In this case the assumption of stationarity of inflow, movement and utilization of water might result in considerable errors when formulating a water management strategy.

The need to study WRS under non-stationary conditions is also caused by the increased anthropogenic impact on natural systems, with possible irreversible consequences, as well as by changes in public opinion concerning human impact on nature and possible climate variations. Thus, the transition from the methodology and strategy of rational water use under stationary conditions to those under non-stationary ones is required, i.e. it is essential to study a WRS as a manageable system and formulate principles of decision-making on its operation and development under non-stationary conditions. It is also necessary to develop a system of mathematical models, following these principles. Such studies should result in the formation of a comprehensive idea on regularities and relevant relations of WRS as natural, technological and socio-economic systems.

2. Aim of the study

The development of well-founded strategies of rational water use under non-stationary conditions is based on the investigation of a WRS as a dynamic manageable system. Some properties of WRS were studied previously under stationary conditions and now it is necessary to study them under non-stationary ones. In the case of stationary conditions some of the characteristics were assumed to be important while others were neglected. Under non-stationary conditions these latter may turn out to be relevant and require careful study. Below some properties of WRS are presented:
- *Dynamic properties of a WRS*. A WRS can be managed, observed, identified. It is also characterized by inertia. These properties describe a possibility to obtain and manage WRS data.
- *Adaptive properties* characterize the ability of a WRS to adapt to unpredicted situations. In addition to the dynamic properties, they include the impact of disturbances in the input data on the WRS (e.g. validity of considering a WRS as a manageable system, stability of its regime and parameters.)
- Structural properties include a hierarchic peculiarity of WRS, a peculiarity of control and a character of water inflow, movement and utilization.
- *Reliability* is the ability of a WRS to supply a required amount of water of standard quality to the user.
- Multipurpose character of water use connected with goals, criteria and estimation for WRS operation.

- Relations between WRS and other systems, and mechanisms of interaction. WRS are active systems.

To select an optimal water use strategy means to select a proper alternative out of many possible ones. Such a selection is carried out using a system of mathematical models, reflecting WRS peculiarities under non-stationary conditions, as well as the procedure of coordinating models' solutions. Development of the models and their coordination represent the main tasks of the study. Such models not only ensure the selection of correct controlling impacts, but also help us to study the above properties. An introduction of non-stationary conditions into models of WRS development and operation means a qualitatively new level of studying water resources, making the investigation of aggregation and approximation even more important. Investigations of non-linear effects of WRS interaction with other natural and socio-economic systems is of great importance at present.

The following key-topics for investigation when solving the above problems can be identified: possibilities of extrapolating the methods of the theory of WRS management under stationary conditions to the case of non-stationarity; the principle of a firm (guaranteed) water yield of the WRS and related to it the problem of its probability. Elaboration of these topics means a reconsideration of the theory of WRS management, based on the investigation of the dynamics of non-stationary processes, forecasting the changes and studying the related uncertainty.

A WRS is an element of the infrastructure of the national economy, affecting its development and the development of society. It coordinates water demands and available water resources. The impact of the WRS on socio-economic development of the whole country and its separate regions, poorly studied under stationary conditions, might turn out to be crucial under non-stationary ones and manifest itself as secondary effects, unpredictable when elaborating a water use strategy. Therefore, WRS should be investigated as natural systems, since their response to non-stationary anthropogenic load is not yet studied. A major task here is to identify the secondary effects and evaluate them.

Models of WRS operation and development under non-stationary conditions are both the means for investigating WRS and elaborating the water use strategy under such conditions. In this paper a model for this purpose is presented. It contains a methodology for decision-making on rational water use and water quality management under risk-prone conditions, based on the concept of firm water yield.

Impacts of disturbances in the input data on the models' performance are very important. Adaptation of the above models, which are multi-extremal even in their stationary deterministic variants, and development of methods for coping with the problems arising when they are applied are also important. Aggregation and decomposition allow the structure of the generalized models to be improved and their efficiency increased.

3. Models of the firm water yield of WRS

The suggested model describes optimal water use and water quality control under risk-prone conditions. The WRS operation is analyzed under stochastic conditions of inflow, movement and distribution of water resources, assigned by a finite set of their possible realizations Ω. Stationarity of these conditions is not assumed, and an approximation of stochastic conditions and processes by a finite set of their realizations is used. The WRS operation in discrete time is simulated, in which changes in the regime are possible only during certain (fixed) time intervals. The calculation interval T in this case is divided into variable intervals of duration h_t, by points $t=0, 1, 2, 3,...,N$, such that $\sum_{t \in T_N} h_t = T$.

The model provides the flow description, in which a WRS is presented as a graph $\Gamma(J,S)$, the geometric configuration of which corresponds to the scheme of the simulated system. A set of vertices J corresponds to the location of water sources, reservoirs, river and canal junctions, water withdrawal and return. A set of arcs S corresponds to water users, river reaches and canals. Water and pollutants in the WRS elements are presented by the corresponding flow. Optimum WRS operation is obtained by determining the optimum flow in the graph.

In this paper we use a schematization of WRS described in [1, 2] with some modifications. These were necessary for introducing as model variables the flows $\tilde{x}, \tilde{u}$, corresponding to the calculated (firm) values of the arc flow and storages (in addition to arc flow and storages x, u presenting actual water and pollutant volumes in the WRS). Flows x are introduced into the model to account for possible deviations in the regime of water use from the one assigned and responsible for the decrease in efficiency of water use. Pollutant flows u are introduced to account for water quality (understood here as a correspondence of chemical and biological characteristics of the aquatic medium to the users' standards). Since water quality management is applied under the same conditions as those for water use (i.e. stochastic, implying the deviation of real water quality characteristics from the assigned ones), flow $\tilde{u}$ is introduced to simulate the assigned values of pollutant volumes. The use of resources other than water and the firm water yield of the WRS elements (like amount of transported loads, flood control measures, etc.) are related to flows $\tilde{x}$ and $\tilde{u}$.

Water users, river reaches and canals are presented as arcs with amplifications. The flow at the beginning of the arc x_{st}^{ω} , simulating the intensity of water supply to the user in a river reach or canal, is connected with the flow at the end of the arc $x_{st}^{k\omega}$, simulating the intensity of water withdrawal, such that

$$x_{st}^{k\omega} = K_{st}^{\omega} x_{st}^{\omega}, \qquad s \in S, \ t \in T_N, \ \omega \in \Omega \qquad (1)$$

with a non-negative amplification coefficient $K_{st}^{\omega} \leq 1$.

Water users' demands for the water amount and constraints on the water discharge are reflected in the model by means of creating upper constraints $\overline{x}_{st}^{\omega}$ and lower constraints $\underline{x}_{st}^{\omega}$ for the arc flow x_{st}^{ω}, which by necessity should lie within the range:

$$\underline{x}_{st}^{\omega} \leq x_{st}^{\omega} \leq \overline{x}_{st}^{\omega}, \qquad s \in S, \ t \in T_N, \ \omega \in \Omega \ . \tag{2}$$

In addition to flows x_{st}^{ω}, which can be different under different realizations of the stochastic conditions ω, the guaranteed flows $\tilde{x}_{st}, s \in S, t \in T_N$ are considered as arcs' inputs. These firm (calculated) flows, having the same values for all $\omega \in \Omega$, are the flows which the elements of the WRS "are expected to obtain". They lie within the range:

$$\underline{\tilde{x}}_{st} \leq \tilde{x}_{st} \leq \overline{\tilde{x}}_{st}, \qquad s \in S, \ t \in T_N \tag{3}$$

with the upper $\overline{\tilde{x}}_{st}$ and the lower $\underline{\tilde{x}}_{st}$ constraints imposed by the physical and technological properties of the water resources used. Assigned flows at the arcs' output are not introduced into the model, since they belong only to the simulated arcs and do not flow out of them (unlike flows, presenting real water movement).

The following flows are also introduced into the model: u_{slt}^{ω}, $s \in S$, $l \in L$, $t \in T_N$, $\omega \in \Omega$. They present pollutants and their firm values are denoted $\tilde{u}_{slt}$, $s \in S$, $l \in L$, $t \in T_N$ where L is a set of pollutants under consideration. A set L is divided into subsets L_υ according to the limiting index $\bigcup_\upsilon L_\upsilon = L$

Adding sets L_υ yields υ-th doses of pollutants with values not exceeding 1 [4]. In terms of the arc flows this means that:

$$\sum_{l \in L_\nu} d_{sl} u_{slt}^{\omega} \leq x_{st}^{\omega}, \qquad s \in S, \ \omega \in \Omega, \ t \in T_N, \tag{4}$$

where d_{sl} is the value, inverse to the maximum permissible concentration (MPC) of the l-th pollutant in the flows x_{st}^{ω} of the s-th arc. Requirements for the quality of water, returned to the WRS by the users, are assigned by the same conditions:

$$\sum_{l \in L_\nu} d_{sl} u_{slt}^{k\omega} \leq x_{st}^{k\omega}, \qquad s \in S, \ \omega \in \Omega, \ t \in T_N \ . \tag{5}$$

Also, lower constraints are imposed on arc flows, presenting pollutants:

$$u_{slt}^{\omega} \geq \underline{u}_{slt}^{\omega}, \quad u_{slt}^{k\omega} \geq \underline{u}_{slt}^{k\omega}, \qquad s \in S, \ \omega \in \Omega, \ t \in T_N \ . \tag{6}$$

They reflect technological peculiarities of the WRS elements' operation, e.g. maximum possible degree of wastewater treatment.

Ranges of possible values of the firm water quality parameters are assigned as constraints, similar to eqs. (4) and (5) for the flows $\tilde{u}_{slt}$, corresponding to the firm amounts of pollutants in the water:

$$\tilde{\underline{x}}_{st} \le \sum_{l \in L_v} \tilde{d}_{slt}\, \tilde{u}_{slt} \le \tilde{\overline{x}}_{st},$$

$$\tilde{\underline{x}}_{st}^{k} \le \sum_{l \in L_v} \tilde{d}_{slt}^{k}\, \tilde{u}_{slt}^{k} \le \tilde{\overline{x}}_{st}^{k}, \qquad s \in S, \quad t \in T_N, \tag{7}$$

where $\tilde{d}_{slt}$ and $\tilde{d}_{slt}^{k}$ are non-negative coefficients of relationships between the planned technologies of water utilization.

Non-linear processes of pollutants' transformation are approximated by linear ones in the model and described by a system of differential equations of the Streeter-Phelps type [5]. Solving this systems of equations, we obtain relationships between the pollutant flows on the arcs' inputs and outputs, forming a set $S_I \subset S$ and presenting river reaches, canals, water users without treatment facilities on the graph $\Gamma(J,S)$:

$$u_{slt}^{k\omega} = \sum_{\gamma \in L} A_{st}^{\gamma l \omega}\, u_{s\gamma t}^{\omega}, \qquad s \in S_I, \tag{8}$$

where $A_{st}^{\gamma l \omega}$ are coefficients, characterizing mutual effects of pollutants and forming a non-singular square matrix $A_{st}^{\omega} = \{A_{st}^{\gamma l \omega}; \gamma, l \in L\}$ of substances' transformation. Its dimensions coincide with those of the set L. A functional relation of the type in eq. (8) is absent between the pollutant flows at the inputs and outputs of the arcs representing water users with the treatment plants and forming a set $S_{II} = S \backslash S_I$, since different values of $u_{slt}^{k\omega}$ can be obtained for the same values of u_{slt}^{ω}, depending on the operation regime of the treatment plants. The flows u_{slt}^{ω} and $u_{slt}^{k\omega}$ of the arcs, belonging to the set S_{II}, are related to each other through operation and maintenance (O&M) costs.

Water sources with unregulated discharge (e.g. river runoff) are represented as flow sources of a given intensity b_{i0t}^{ω} and b_{ilt}^{ω}, $l \in L$, located in the vertices $i \in J$.

Water sources with regulated discharge (e.g. interbasin water transfer systems) are presented in the model as fragments $\Gamma(J,S)$, containing additionally introduced vertices. Two arcs, s_1 and s_2, go out of each of them. Their amplification coefficients are $k_{s2t}^{\omega} = 0$ and $k_{s1t}^{\omega} = 1$, respectively. Flows x_{s1t}^{ω}, u_{s1lt}^{ω}, $l \in L$ of the arc s_1 simulate the intensity of water and pollutants inflow from the water source into the WRS. Flows x_{s2t}^{ω}, u_{s2lt}^{ω}, $l \in L$ of the arc s_2 simulate a non-utilized capacity of the source. The flow sources with given intensities b_{i0t}^{ω}, b_{ilt}^{ω}, $l \in L$, correspond to the maximum amounts of water and pollutants that can be obtained from the source during a time unit. These sources are located in the introduced vertices. Ranges of the possible values x_{s1t}^{ω}, x_{s2t}^{ω}

are assigned by the conditions of the type given in eq. (4), and values u^{ω}_{s1lt} , u^{ω}_{s2lt} - by the conditions of the type in eq. (2). O&M costs of such sources belong to the arcs s_1. Impacts of the O&M costs for the arcs s_2 are assumed to be zero.

Amounts of water and pollutants to be withdrawn from the source are different under different realizations of the stochastic conditions ω. At the same time it is necessary to plan the water demands and other water related resources. Therefore, the source should be characterized by its firm water yield. In the model the calculated flows $\tilde{x}_{s1t}$ and $\tilde{u}_{s1lt}$, $l \in L$, correspond to the firm (guaranteed) water yield. In order to unify the description of arcs, the calculated flows $\tilde{x}_{s2t}$ and $\tilde{u}_{s2lt}$ are introduced, which does not change anything in the model since the cost function of the arc s_2 is zero. Flows $\tilde{x}_{s1t}$ lie in the intervals defined by the conditions in eq. (3), and the flow $\tilde{u}_{s1lt}$ - by those in eq. (7).

Reservoirs are considered in combination with water users located on their banks. They are represented as storages in the vertices $i \in J$, corresponding to the amount of water in the reservoirs u^{ω}_{ilt} , and $l \in L$ – to the amount of pollutants in the water. Requirements of water users and properties of the water bodies themselves create a set of possible storage values:

$$\underline{x}^{\omega}_{it} \le x^{\omega}_{it} \le \overline{x}^{\omega}_{it} ; \tag{9}$$

$$\sum_{l \in L_v} d_{il} u^{\omega}_{ilt} \le x^{\omega}_{it}; \qquad i \in J, t \in T_N, \omega \in \Omega, l \in L \tag{10}$$

$$\underline{u}^{\omega}_{ilt} \le u^{\omega}_{ilt}; \tag{11}$$

Reservoir water losses are approximated in a linear form as:

$$\delta x^{\omega}_{it} = \gamma^{\omega}_{it} x^{\omega}_{it}, \qquad i \in J, \quad t \in T_N, \quad \omega \in \Omega \tag{12}$$

with the coefficient of losses γ^{ω}_{it} varying from 0 to 1. It follows from the solution of the Streeter-Phelps equations that pollutant storages answer the requirements of the system of linear equations:

$$u^{\omega}_{il,t+1} = \sum_{\gamma \in L} A^{\gamma l \omega}_{it} u^{\omega}_{i\gamma t} + h_t v^{\omega}_{ilt}, \qquad i \in J \tag{13}$$

where coefficients $A^{\gamma l \omega}_{it}$ are similar to $A^{\gamma l \omega}_{st}$ in eq. (8); v^{ω}_{ilt} is the flow of the l-th pollutant, entering the i-th storage or leaving it.

Besides the storages x^{ω}_{it} and u^{ω}_{ilt} , which can have different values for different $\omega \in \Omega$, calculated (firm) storages $\tilde{x}_{it}$ and $\tilde{u}_{ilt}$, $l \in L$, which are independent on ω , are

164

introduced into the model. Values of the storages $\tilde{x}_{it}$, simulating the amount of water in the reservoir, lie within the range:

$$\underline{\tilde{x}}_{it} \leq \tilde{x}_{it} \leq \overline{\tilde{x}}_{it}, \qquad i \in J, \quad t \in T_N \tag{14}$$

with the lower ($\underline{\tilde{x}}_{it}$) and the upper ($\overline{\tilde{x}}_{it}$) constraints largely determined by the use of resources other than water. Ranges of storage values $\tilde{u}_{ilt}$, $l \in L$, corresponding to the amount of pollutants in the reservoir water are assigned similarly to eq. (7) as:

$$\underline{\tilde{x}}_{it} \leq \sum_{l \in L_v} \tilde{d}_{ilt} \, \tilde{u}_{ilt} \leq \overline{\tilde{x}}_{it}, \tag{15}$$

The model takes into account <u>the law of conservation of mass</u> of water and pollutants. A system of equations for the flows continuity in the graph's vertices follows this law:

$$x_{i,t+1}^{\omega} - x_{it}^{\omega} = h_t \left[\sum_{s \in S_i^+} x_{st}^{k\omega} - \sum_{s \in S_i^-} x_{st}^{\omega} - \delta x_{it}^{\omega} + b_{ilt}^{\omega} \right], \tag{16}$$

$$v_{ilt}^{\omega} = \sum_{s \in S_i^+} u_{slt}^{k\omega} - \sum_{s \in S_i^-} u_{slt}^{\omega} + b_{ilt}^{\omega}, \tag{17}$$

where S_i^+ is the assembly of arcs going into vertex i , S_i^- - arcs leaving this vertex. Some pollutants find their way into river reaches and canals together with water. (Their concentration in this case is equal to that at WRS sites). Therefore, the following conditions should be fulfilled in vertices $i \in J$:

$$u_{slt}^{\omega} \, x_{\tilde{s}t}^{\omega} = u_{\tilde{s}lt}^{\omega} \, x_{st}^{\omega}, \qquad s, \tilde{s} \in S_i^-, \tag{18}$$

which means the coincidence of the flow compositions of the arcs, leaving one and the same vertex.

The cost function is based on the assumption that the effects of water use by different WRS elements can be expressed in the same units, (e.g. rubles). It allows calculating WRS operation under one of the ω realizations applying one defined criterion f^{ω} (costs), which depends on the projected ($\tilde{x}, \tilde{u}$) and actual (x^{ω}, u^{ω}) regimes. Besides, the cost for the whole system is assumed to be the ensemble of the costs of its elements:

$$f^{\omega} \left(\tilde{x}, \tilde{u}, x^{\omega}, u^{\omega} \right) = \sum_{r \in R} f_r^{\omega} \left(\tilde{x}_r, \tilde{u}_r, x_r^{\omega}, u_r^{\omega} \right) \tag{19}$$

where $R=J$ and S is a set of WRS elements.

Vectors $\tilde{x} = \{\tilde{x}_r; r \in R\} = \{\tilde{x}_{rt}; r \in R, t \in T_N\}$ and $x^\omega = \{x_r^\omega; r \in R\} =$
$= \{x_{rt}^\omega; r \in R, t \in T_N\}$ join the arc flows and storages corresponding to the regime of
water use in the WRS. Vectors $\tilde{u} = \{\tilde{u}_r; r \in R\} = \{\tilde{u}_{rlt}, \tilde{u}_{rlt}^k; r \in R, l \in L\}$ and
$u^\omega = \{u_r^\omega; r \in R\} = \{u_{rlt}^\omega, u_{rlt}^{k\omega}; r \in R, l \in L, t \in T_N\}$ join the arc flows and storage
simulating the amount of pollutants in the WRS.

A regime, minimizing the total expenses under any realization of the stochastic
conditions ω, is preferable for the WRS. That is why it is expedient to adopt the
mathematical expectation of the expenses as the objective function:

$$f(\tilde{x},\tilde{u},x,u) = \mathop{M}_{\omega} f^\omega(\tilde{x},\tilde{u},x^\omega,u^\omega) \tag{20}$$

where $x = \{x^\omega; \omega \in \Omega\}$ and $u = \{u^\omega, \omega \in \Omega\}$.

A set of possible stochastic conditions Ω is a finite number of implementations:

$$f(\tilde{x},\tilde{u},x,u) = \sum_{\omega \in \Omega} P^\omega f^\omega(\tilde{x},\tilde{u},x^\omega,u^\omega) = \sum_{r \in R}\sum_{\omega \in \Omega} P^\omega f_r^\omega(\tilde{x}_r,\tilde{u}_r,x_r^\omega,u_r^\omega) =$$
$$= \sum_{r \in R} f_r(\tilde{x}_r,\tilde{u}_r,x_r,u_r) \tag{21}$$

where P^ω is the probability of the realization ω and f_r is the mathematical expectation of
the expenses of the r-th element.

4. Problems of the optimal operation of WRS

The optimal vector of the arc flows and storages $\tilde{x}^0, \tilde{u}^0, x^0, u^0$ simulating water and
pollutant movement corresponds to the optimal operation regime of a WRS, with regard
to both quantity and quality of the water resources. This vector represents the solution
of the problem (called "A" below) minimizing the cost function, determined by the
equality of expenses eq. (21) on the set G_A, defined by:

a   rules eqs. (1), (8), (12), (13) for the arc flows and storage transformations
simulating water and pollutant movement in the WRS;

b   eqs. (16) and (17) describing the law of conservation of mass of water and
pollutants in the WRS;

c   relationship eq. (18) describing the interrelationship of water and pollutant flows in
the WRS;

d   constraints eqs. (2) to (7), (9) to (11), (14) and (15) for the arc flows and storages
on the graph $\Gamma(J,S)$ simulating the projected and actual amounts of water and
pollutants in the WRS elements.

The problem "A" is a task of a two-step stochastic programming, in which the
strategic variables on the first step, selected for the unknown realizations of the

stochastic conditions ω, are assigned the values of the arc flows and storages $\tilde{x}, \tilde{u}$ simulating the calculated amounts of water and pollutants. The tactical variables on the second step, selected for the known realizations of the stochastic conditions, are the arc flows and storages x^{ω}, u^{ω} corresponding to the amounts of water and pollutants entering the WRS element and leaving it.

Solving the problem "A" implies minimizing the non-convex function $f(\tilde{x}, \tilde{u}, x, u)$ on a non-convex set G_A. The latter is non-convex due to the non-linearity of the constraints eq. (18). The task is multi-extremal, i.e. having local minima. There are no general methods for solving such problems. However, a specific character of the problem "A" allows applying a solution with an error not exceeding the assigned one. The specific character of this problem consists in the generalized separability of the objective function eq. (21) /k- separability/, i.e. the task function can be presented as an ensemble of functions, each of them depending on a small number (less than k) of variables. The separability of the problem "A" also shows up in the bilinearity of the constraints eq. (18). Here the determination of the optimal vector in the problem "A" is reduced to solving a finite set of evaluative tasks of convex programming, formed by means of determining the convex covers for the separate items of the objective function and the constraints on the system of constricting rectangular sets [3]. This method refines the scheme of branches and boundaries when solving multi-extremal tasks.

In addition to the optimum values of the firm water quality and quantity, the solution of the problem "A" $(\tilde{x}^0, \tilde{u}^0, x^0, u^0)$ includes a determination of the optimal (calculated) probability. The latter is understood as the possibility that WRS operation regimes belong to the set of the optimal ones. Favorable (optimal) regimes are the ones ensuring the use of a smaller amount of water with a lower pollutant content compared to the assigned values. The optimal probability is the one at which a WRS is operated with the least possible expense.

Since the optimal vector $\tilde{x}^0, \tilde{u}^0, x^0, u^0$ of the problem "A" contains the optimal values of the arc flows and storages for various realizations of the stochastic conditions ω, the distribution of the stochastic values $x_{rt}^{\omega 0}$ and $u_{rlt}^{\omega 0}$ corresponds to this vector. Knowing this distribution and the calculated flows and storages $\tilde{x}_{rt}^0$, $\tilde{u}_{rlt}^0$, we can determine the optimal (calculated) probabilities for different types of water use as the probabilities of "non-violation" of the favourable regimes. For instance, for the favourable regimes in which $x_{rt}^{\omega 0} \geq \tilde{x}_{rt}^0$ and $u_{rlt}^{\omega 0} \leq \tilde{u}_{rlt}^0$, the optimum probabilities for the absence of the deficit in the required water amounts with the assigned quality for the users, river reaches and canals are:

$$P_r^0 = \sum_{\omega \in \left\{ \omega \in \Omega \mid \sum_{t \in T_N} h_t \left[x_{rt}^{\omega 0} - \tilde{x}_{rt}^0 \right] \geq 0, \ \sum_{t \in T_N} h_t \left[u_{rlt}^{\omega 0} - \tilde{u}_{rlt}^0 \right] \leq 0 \right\}} P^{\omega} \qquad (22)$$

For the reservoirs this probability is:

$$P_r^0 = \sum_{\omega \in \left\{\omega \in \Omega \mid \left[x_{rt}^{\omega 0} - \tilde{x}_{rt}^0 \geq 0, \left[u_{rlt}^{\omega 0} - \tilde{u}_{rlt}^0 \leq 0, t \in T_N \right\}} P^{\omega} \qquad (23)$$

The optimum probability for a number of continuous periods is as follows:

$$P_r^n = \sum_{\omega \in \Omega} P^{\omega} \frac{1}{N} \sum_{t \in \left\{t \in T_N \mid x_{rt}^{\omega 0} - \tilde{x}_{rt}^0 \geq 0, u_{rlt}^{\omega 0} - \tilde{u}_{rlt}^0 \leq 0 \right\}} \delta(t) \qquad (24)$$

where $\delta(t) = 1$ if $t > 0$, $\delta(t) = 0$ if $t \leq 0$.

The optimum probability for the duration of the continuous periods of water supply and constant quality of water resources is calculated as:

$$P_r^R = \sum_{\omega \in \Omega} P^{\omega} \frac{1}{T} \sum_{t \in \left\{t \in T_N \mid x_{rt}^{\omega 0} - \tilde{x}_{rt}^0 \geq 0, u_{rlt}^{\omega 0} - \tilde{u}_{rlt}^0 \leq 0 \right\}} h_t \qquad (25)$$

It should be noted that different forms of the calculated probabilities correspond to different types of water use. A variety of probabilities is obtained by solving the problem "A".

An assignment of the probabilities of water use differing from the optimal ones cannot guarantee a decrease in the management expenses because this means that the conditions, restricting the permissible set of the problem "A", do not decrease the optimal value. Therefore, these restrictions are added to the constraints defining the set G_A. If $P_t < P_r^0$, the costs increase due to more frequent expensive adjustments than usually needed. These adjustments are necessary due to the deviations in the amount of water utilized from the guaranteed amount. If $P_t > P_r^0$, the expenses grow due to an incomplete account for all the possibilities of water use when projecting the WRS.

5. Conclusions

This paper has presented different strategies of WRS management under non-stationary conditions. The solution of this problem is connected with the solution of individual problems presented in the paper. The major tasks are identified among them, including the development of mathematical models for studying WRS properties and elaboration of a strategy for rational water use under non-stationary conditions. A mathematical model of the WRS firm water yield and its probability has been presented. It needs to be included in a system of models describing the development and operation of WRS under non-stationary conditions. The model contains the methodology for decision-making concerning rational water use and water quality management under risk-prone conditions.

168

6. References

1. Aloev, T.B., Kapusta, A.E. and Khranovich, I.L. (1987) Flow models for the selection of optimal parameters of water resource systems. *Vodnye Resursy* **1,** 20-34 (in Russian).
2. Kocharian, A.G. and Khranovich, I.L. (1986) A mathematical model of planning optimum parameters of water resource systems, water allocation and control of surface water quality. In : *"Systems Analysis Applied to Water and Related Land Ressources." Proc. IFAC Conf., Lisbon, Portugal*, Pergamon Press, Oxford, 75-79.
3. Lazebnik, A.I., Khanovich, I.L. and Tsallagova, O.N. (1981) Generalized separable problems and their application. *Avtomatika i Telemekhanika* **8,** 107-118 (in Russian).
4. *Ukazanija po Metodike Rascheta i Razbavlenija Stochnykh Vod v Rekakh, Ozerakh i Vodokhranilischakh. (Instructions on Methods of Calculating and Mixing of Wastewater in Rivers, Lakes and Reservoirs)* (1971) VODGEO Publ Moscow, 224pp (in Russian).
5. Vavilin, V.A. and Tsitkin, M.Yu (1977) Mathematical modelling and water quality management. *Vodnye Resursy* **5,** 114-132 (in Russian).

Interactive River-Aquifer Simulation and Stochastic Analyses for Predicting and Evaluating the Ecologic Impacts of Alternative Land and Water Management Policies

Daniel P. Loucks
School of Civil and Environmental Engineering
Cornell University
Ithaca, NY 14853 USA

1. Overview

This paper describes an Interactive River-Aquifer Simulation program for estimating time series of flows, storage volumes, water qualities and hydroelectric power and energy produced and consumed from pumping throughout a surface water, or an interdependent surface and groundwater, system. Alternative land and water management policies and practices can be simulated and evaluated. A stochastic analysis component of this program permits the prediction of the ecological impacts of each management policy.

IRAS has been developed primarily to assist those interested in evaluating the performance or impacts of alternative management and operating policies of regional water resource systems. Such systems can include multiple interconnected rivers interacting with multiple aquifers serving large regions, or they could include only a portion of a single river or stream in a small watershed. The user defines and has control over the spatial as well as the temporal resolution of the system being simulated.

IRAS is data driven. The input data define the system configuration, the system components, their design parameters, and how each of those components operate, either independently or interdependently. The complexity of the system design and operation, and its spatial and temporal resolutions, will determine the amount and state of input data needed. The user can readily change all system design and operating data. This facilitates sensitivity analyses. Both the input and output data can be expressed in any units desired by the user.

The systems to be simulated using IRAS must be represented by a network of connected nodes and links. The user must draw this network into the graphics terminal. This network (for the PC version of IRAS) can have up to 400 nodes and 400 links. The nodes of the network can represent aquifers, gage sites, consumption sites, natural lakes, reservoirs, wetlands, confluences

M.V. Bolgov et al. (eds.), Hydrological Models for Environmental Management, 169–194.

and diversions. A single node may be a feasible combination of any of these node states.

The links of the network can be uni-directional (flow only in one direction) or bi-directional (flows in either direction, as for pumped storage operations or for flows dependent on changing surface elevation or pressure head differences). Links can represent river reaches, diversions, and the transfer of water among aquifers and/or wetlands and the surface water system. Any river reach or diversion may contain a hydroelectric power plant, or pumping station (e.g., for pumped storage plants or for aquifers subjected to pumping and/or artificial recharge). Flow routing in uni-directional links is also possible.

IRAS can simulate independent or interdependent water quality constituents. The user must define the water quality constituents to be simulated, the growth, decay, and transformation rate constants pertaining to those constituents, and other parameters identifying the form of the water quality model used in the water quality portion of the simulation program.

The IRAS program is capable of simulating water resource systems over multiple within-year time periods for any portion of one year or for any number of multiple years. Each within-year period is divided into multiple simulation time steps. The user must define the number of within-year periods to be simulated and the duration of each of them (usually a day or longer). Input and output data pertain to each within-year period. Any number of years, or portion of a single year, may be included in any simulation run.

Applications of IRAS may simply involve the prediction, over time and space, of the values, or of user-defined functions of the values, of the simulated variables for various hydrologic and water quality inputs. These outputs at a given node or link can be displayed as time-series plots. Outputs of a sequence of connected nodes or links over space can be displayed as space-series plots, one for each successive within-year time period. Green, yellow and red colors can be assigned to various ranges of variable values judged by the user to be satisfactory, marginal, or unsatisfactory, respectively. Then these color-coded representations of the ranges of these selected variable values can be displayed geographically on the schematic network or map, if defined, over successive time periods. This provides a means of identifying relatively quickly the locations and times where the system may be stressed.

Applications can also include the prediction and display of statistical measures and the probability distributions of possible durations and extents of failures, if any, from meeting user-defined flow, hydroelectric energy, storage volume, or water quality concentration targets at various sites within the system. Simulations using IRAS can help identify the range and likelihood of various water quantity, water quality, hydropower production and pumping energy consumption, and ecological impacts that may be associated with any particular land and water management and use policy or practice. The results of multiple simulations under different assumptions can be compared.

2. Introduction

Considerable investments have been made in the design and operation of structures and facilities used to control water quantity or quality within many river basins in the world. More are being planned for future installation and operation. These facilities have multiple impacts over both time and space. Some of these impacts may be beneficial, others may not. Simulation models can help identify these impacts and the extent they might be altered by changes in system design or operating policy.

Computer simulation of water resource systems is one of the best ways for evaluating alternative design parameters and operating policies of system components, and for estimating the likelihood of possible impacts resulting from these alternative system designs and operating policies. There is no better way of obtaining this information, short of actually implementing a particular system and observing its performance. The latter option is clearly expensive and risky, especially if the system doesn't perform very well.

Given any set of assumptions regarding system design and operation, simulation models can provide a way of keeping track of where water is, where it goes, and what its quality is, for a fixed sequence of hydrologic and wasteload inputs during a fixed sequence of time periods. If applicable, the amount of power and energy generated and used for pumping can be determined. If the hydrologic and wasteload inputs are representative of what might occur in the future, the simulation results should be indicative of what one would expect to observe in the future, at least in a statistical sense. Through multiple simulations, individuals can test, modify, and evaluate various designs and operating policies in a systematic search for the ones that they believe perform best.

It is obvious that the results of any computer simulation are always very dependent on the assumptions incorporated into the simulation model. The assumptions that must be incorporated into any simulation program are numerous, and are often based on unknown or uncertain information. These assumptions typically include future hydrology, future user demand and consumption rates, changing facility design and operation, and numerous values of parameters of functions that define how the system works, e.g., the production of hydroelectric energy, the flow of groundwater, and the fate and

172

transport of pollutants. It is impossible to know all of these assumptions with precision.

It is not the purpose of any simulation model to identify which credible set of assumptions is best. Simulation models can, however, be used to estimate what, when, and where something may happen, given any set of assumptions. These models can be used to test how sensitive the performance of any particular system may be to various uncertain inputs and parameter values. Through simulation, individuals can identify which assumptions substantially affect the performance of the system and hence which assumptions may be worthy of additional study and examination. After numerous simulation runs under different hydrologic conditions, system designs, and operating policies, humans (not models) can better judge which designs and operating policies may provide the greatest expected benefits or have the lowest expected adverse risks under the most plausible set of assumptions.

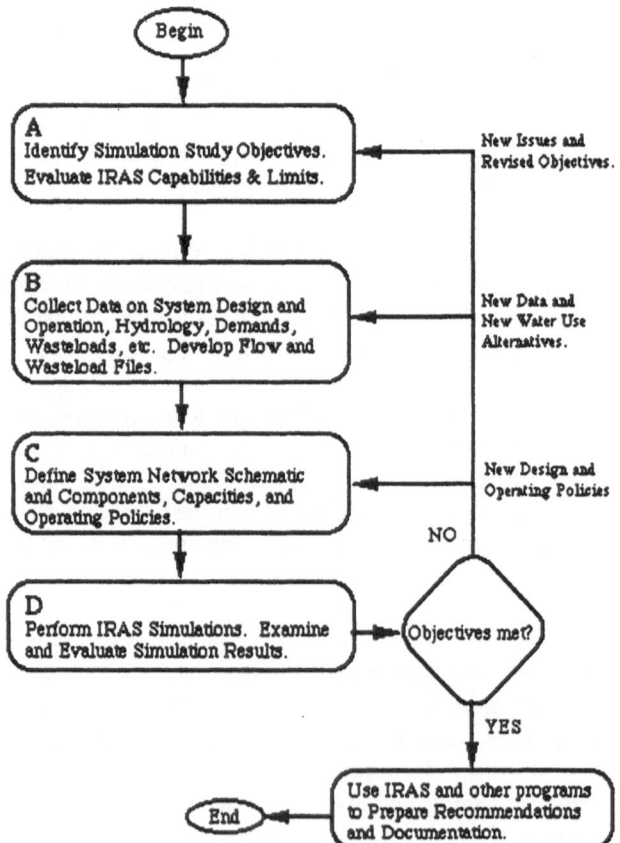

Figure 1. The interactive river-aquifer simulation (IRAS) process.

3. IRAS Simulation Process

Figure 1 outlines the IRAS simulation process. The process is similar to that used with any water resources planning model. How well each step of the process is performed influences the applicability and reliability of the simulation results and the necessary number of iterations through the simulation process.

Beginning in Box A of Figure 1, it is important to note that a potential IRAS user should have a thorough understanding of both the objectives of the simulation study and the capabilities and limitations of the IRAS model.

IRAS can be used to evaluate the performance of any specific system configuration and set of operating policies, but it cannot automatically identify the more preferred design or operating policy alternatives. IRAS does not contain a system optimization capability that eliminates the need to specify operating policies everywhere choices are to be made. Users must identify each alternative design or operating policy that is to be simulated. The results can be compared to those of other simulations. Careful consideration of simulation study objectives and IRAS' capabilities and limitations can reduce the time, data, and expense required for these multiple simulations and comparisons.

Since IRAS can perform a reasonably detailed simulation of the operation of a water resource system, it is necessary to collect and analyze data characterizing the system's design and operation and to convert it to a form suitable for entry into IRAS. This process of data collection and analysis and conversion is likely to continue as long as IRAS is being used to simulate the system. All the steps between Boxes B and D in Figure 1 are likely to be repeated many times whenever IRAS is used in a study examining and evaluating design and operating policy alternatives.

It is generally most efficient to perform the tasks outlined in Box B as completely as possible early in the study. Even so, it is possible that supplemental data will have to be obtained and analyzed throughout the study.

The data collection activities described in Box B of Figure 1 must be performed with care. As will be explained in more detail later in this manual, IRAS depends upon the development of a set of "uncontrolled" streamflow records at so-called gage sites, groundwater recharge estimates (when applicable) and "gage multipliers" that spatially interpolate or extrapolate the gage-site estimates to non-gauged sites.

The development of these uncontrolled flow data, evaporation and seepage losses, and aquifer natural recharge rates, may well be among the most difficult and labor intensive activities required when applying IRAS to an actual system. Errors in these input data will certainly affect the accuracy of the simulation output data.

The IRAS program was developed primarily to increase the efficiency of performing the tasks outlined in Boxes C and D in Figure 1. If tasks A and B are performed well, then IRAS can decrease the effort associated with, and significantly increase the effectiveness of, a regional water resources study.

The data presentation and statistical analysis procedures incorporated into IRAS are designed to help users identify and understand important simulation results. These display capabilities include time-series and space-series plots of user-selected simulated variables and user-defined functions of these variables, as well as color-coded static and dynamic plots (over digitized maps) of system performance measures. The statistical displays include calculated reliabilities, resiliences, and vulnerabilities as well as unconditional and conditional probability distributions of these measures at user selected sites.

4. IRAS' Interactive Interface

The IRAS simulation program menu-driven interface has been developed to give the user control over the operations of the program. These menus provide the user with a number of options that can be implemented. The user can control the sequence of program operations with a pointing device (such as a mouse or pen and tablet) and a keyboard. The inputting of data is done in a manner that facilitates the detection of mistakes and allows for easy modification of those data for sensitivity analyses. Users can display the simulation output data in a variety of ways, and each way is intended to improve the users level of understanding.

The interface built into IRAS has been designed to facilitate user interaction in:

- creating or editing a digitized map image of the region or river basin for geographic referencing when displaying spatial data
- inputting, retrieving and editing system configuration and design data
- defining a water quality model for user-defined quality constituents
- defining alternative system component operating policies
- operating the simulation program
- displaying, editing, and printing/plotting the simulation inputs, results and analyses

- performing and displaying statistical analyses of simulation results.

Figure 2 illustrates the interface of the IRAS program. The node-link network is drawn over a map of the region, and each node and link have data pertinent to its operation. Photographs and videoclips of the region being modeled can also be displayed, as desired.

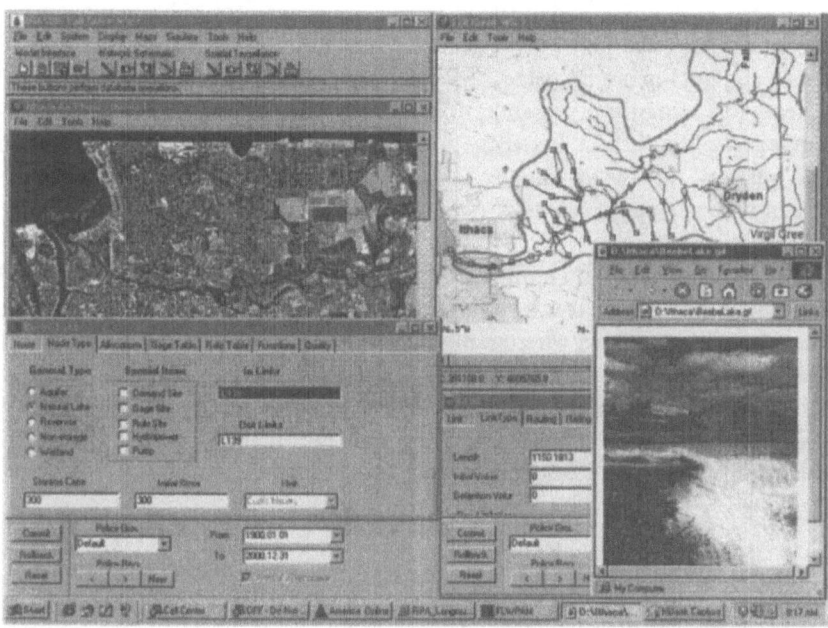

Figure 2. Interface of IRAS showing main menu items of water modeling framework in upper left and various windows of the IRAS program into which a node-link network is drawn, and data are entered defining what each node and link is, and how each node and link operate. In the lower left, a video clip taken at a node site is being displayed.

5. Using IRAS for Planning and Management

There are many kinds of models that can be used to study various water resources planning and management issues. Some of these models are designed to examine in considerable detail the hydrologic or hydraulic processes that take place in watersheds, in water bodies, or in facilities built to remove wastes from water or wastewater. Other models are designed to identify and evaluate potentially attractive facility and management alternatives prior to their detailed design. The IRAS computer simulation program described in this document is designed to assist those responsible for planning and management. Using continually updated information, it could also be used for forecasting possible outcomes of current operations. It is not a program suitable for detailed hydraulic design.

IRAS is a relatively simple simulation model. Its simplicity reduces the input data required for simulation as well as what can be obtained from the results. Its simulation is based on mass balances of quantity and quality constituents, taking into account flow routing, seepage, evaporation, and consumption, as applicable. IRAS is a one-dimensional model. Each natural lake, reservoir, aquifer and wetland-area component (or sub-component) is modeled as a storage unit - a simple bath tub, but one whose distributions of quantity and quality are not necessarily uniform. Within the accuracy provided by these very simplifying assumptions, IRAS attempts to address problems involving the interaction between ground and surface waters, and between water quantity and water quality. These processes typically involve quite different time and space scales.

IRAS is generic, i.e., it is designed to be applicable to many different water resource systems. It is a simulation model that may be useful in planning or management situations where a more specific or more detailed model of the water resource system is not available. IRAS is intended for use as a first step, for helping one determine whether or not a more detailed simulation is warranted. If IRAS cannot provide the information needed, then an alternative, often more site-specific, model should be used.

6.   Capabilities and Limitations

IRAS has been developed primarily to assist those interested in simulating the spatial and temporal distributions of flows, storage volumes, water quality, and hydropower production and/or energy consumption in water resource systems that can include river or streams, diversion canals, lakes, reservoirs, wetlands and aquifers, together with multiple users. The flow and storage quantity and quality variable values are dependent on the designs and operating policies being simulated as well as on the water quantity and quality inputs and user-defined model parameters. The systems being simulated can include interconnected multiple rivers and aquifers serving large regions or only a portion of a single river or stream in a small watershed.

The systems to be simulated using IRAS can include up to 400 stream and river reaches (including diversions such as canals or pipelines), and up to 400 sites that can represent any feasible combination of aquifers, gage and monitoring sites, consumption sites, natural lakes, wetlands, reservoirs, confluences and diversion or withdrawal sites, hydropower and pumping sites, and wastewater effluent discharge sites. Flows can be uni-directional (one way flow) or bi-directional (flow in either direction). Flows can also be in both directions (e.g., as in some pumped storage systems).

IRAS can simulate independent or interdependent water quality constituents. The user must define the water quality constituents to be simulated, the growth, decay, and transformation rate constants, and other parameters used in the water quality portion of the model. Some assumptions built into the program limit the states of systems and constituents that are appropriate for water quality simulation. The quality model considers only first-order kinetics of constituent growth, decay, and transformation and only one-dimensional (longitudinal) advection and dispersion. All river reaches, lakes and reservoirs are assumed to be vertically and laterally well-mixed. Hence the predicted average values of each quality constituent provided by this simulation program will not reflect variation over two or three-dimensional space.

The IRAS program is not capable of simulating flood or other events requiring hydraulic flow routing. It is also not capable of explicitly simulating within-day changes in water temperature and light. In addition, IRAS cannot simulate policies that change over multiple years. However, if the demand for water at any site, such as an irrigation area, depends on the local rainfall, streamflow, or storage at that site, then demand-driven targets can be defined and the actual water allocations to be made to satisfy these targets can vary over time from year to year.

7. Spatial and Temporal Resolution

The user of IRAS specifies and controls the spatial and temporal resolutions of the desired simulation. IRAS can simulate a single reach of a stream, or multiple interconnected river basins draining multiple countries. The size of the region and the detail being modeled within the region is entirely up to the user. The program is capable of simulating such water resource systems over multiple within-year time periods over a succession of years. Users have control over the number of within-year periods and the duration (a day or longer) of each within-year period.

The program divides each within-year period into 12 simulation time steps. The program computes the flows, storage volumes, energy, and water quality constituents throughout the system in each simulation time step. All input and output data pertain to initial, average, or final conditions for each within-year period, as appropriate. The output of IRAS includes the initial and final storage volumes and the average flow, quality, energy and power conditions for each of the user-specified within-year periods in each year of the simulation.

178

8.   System Configuration and Operation

IRAS has been designed to be data-driven. Input data define the system, its configuration, its components and the design parameters of those components, and how each of those components operate, either independently or interdependently. All system design and operating assumptions contained in the model are defined by the input data provided by the user and can be readily changed by the user. The program provides considerable flexibility regarding the system configuration and operation. This flexibility is one of the primary advantages IRAS has as a planning tool. The program allows various facilities and components of a water resource system to be modeled with varying degrees of detail. Users can develop and include data that characterize each facility at its appropriate level of detail.

Operating policies are needed wherever choices must be made. These include all reservoir sites where releases must be specified, node sites where water can be allocated to multiple outgoing links or to a single outgoing diversion link, and groundwater well sites where pumping may occur, either for withdrawals or for artificial recharge. Decisions regarding how much water goes where at each decision point in any simulation time step can be based on the demand for water at various consumption or demand target sites and on the available supply at that point.

8.1   ALLOCATION AND CONSUMPTION FUNCTIONS

In supply driven cases, allocation functions can be defined. These functions indicate the amount of water entering each link, given the total outflow. Alternatively, water may enter each link to meet user-defined demand targets downstream. (In the first case the node's outflow is pushed into each link, and in the latter case the water is pulled into each link.)

These allocations are defined as piece-wise linear functions of the total node outflow, as illustrated in Figure 3. The allocation functions may differ in each within-year period, but not from year to year.

8.2   RESERVOIR RELEASE POLICIES

For each surface water reservoir node, the user must enter data into the node's data dialog boxes that define its operation in each within-year time period. A reservoir operating policy can be based on the available supply and known demands, or on downstream demand deficits, or both.

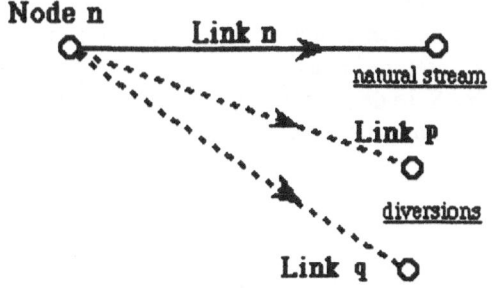

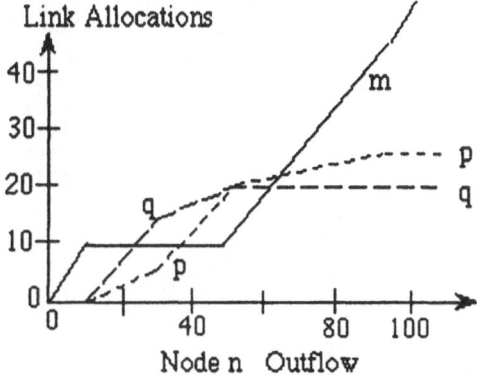

Node n Outflow	Outflow Allocations to		
	Link m	Link p	Link q
0	0	0	0
10	10	0	0
30	10	5	15
50	10	20	20
90	45	25	20

Figure 3. Supply-driven link allocation functions for the two diversion links (dashed lines) and a main channel link carrying water from the node n. The total amount allocated depends on the total amount of water available to allocate.

The options to be described for managing the spatial and temporal distributions of flows, storage volumes, water qualities, energy, and power that are provided in IRAS provide considerable flexibility, but in some cases they may have to be modified to meet specific operating requirements or needs of specific systems. In these cases separate subroutines can be written and added to IRAS to modify the operating options included within IRAS. A reservoir release policy based on storage volume zones requires the user to define the number of such zones, and the storage volume boundaries of each of those

180

zones, for each within-year period. In addition, the releases associated with the storage volumes in each of the storage volume zones must be defined. Figure 4 illustrates a reservoir release rule based on storage zones, and the data required, by IRAS, to define the zones and the releases associated with storage volumes in those zones. Hence if a reservoir release policy is defined, the user must enter:

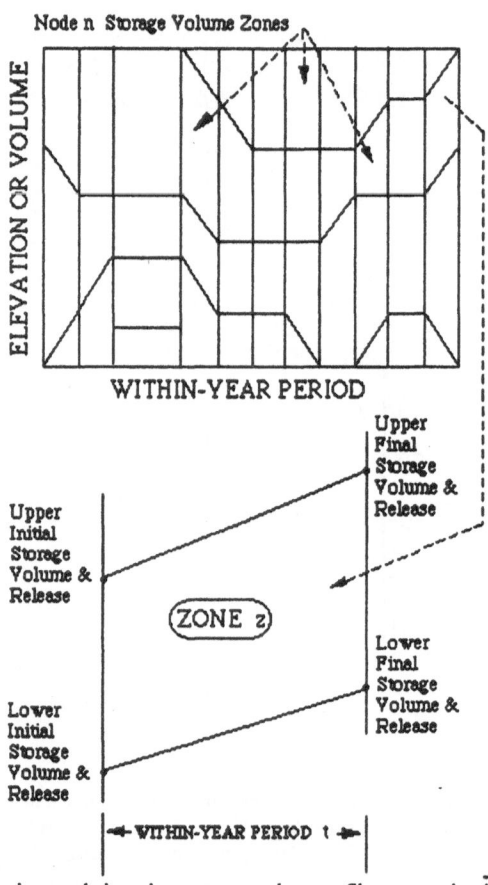

Figure 4. Reservoir release rule based on storage volumes. Shown are the data (the volumes and corresponding releases at the corners of each zone) required to define the storage zones and their releases. This information is needed at each reservoir release rule site.

- The number of zones and the storage volumes and releases associated with each release zone if the reservoir is at a reservoir release rule site.

- Target storage volume levels as a function of total reservoir group storage volume if the reservoir is in a group of multiple reservoirs and not at a reservoir release rule site.

For multiple reservoirs being operated interdependently as a group, the reservoir release rule will be a function of the total storage in the group of reservoirs. If the reservoir group release rule is at a reservoir site, usually the most downstream reservoir of the group, the release from that reservoir will be determined exactly as described above for a single independent reservoir. The release from that release rule site reservoir will be determined from the release rule, or the minimum required release, whichever is greater. If the available storage volume and inflow are insufficient to make this release, then the total amount of water in the reservoir will be released.

The releases from all reservoirs in a reservoir group not located at the release rule site will be determined from user-defined reservoir storage target or balancing functions. These balancing functions, as illustrated in Figure 5, must be defined for each within-year period. They indicate the desired or target storage in each upstream reservoir of the group as a function of the current total reservoir storage in the entire group. The simulation model will attempt to keep all reservoirs in balance during the simulation by releasing in each simulation time step any water in excess of the target storage volume.

8.3 RESERVOIR RELEASES BASED ON DEMAND

Reservoir release policies can also be based on demand deficits at designated demand nodes. Releases based on demand deficits, prorated over the remainder of the within-year period, are added to the releases based on supply-driven release rules and balancing functions.

8.4 GROUNDWATER PUMPING

If pumping is implemented to withdraw or add water, the flow pumping policies must be defined. These pumped flows can be a function of current storage volumes in the aquifer or wetland nodes and the water available (initial storage, if any, less any seepage and evaporation losses plus inflow) in the surface-water nodes at each of the ends of the groundwater or wetland-area link.

RESERVOIR GROUPS:

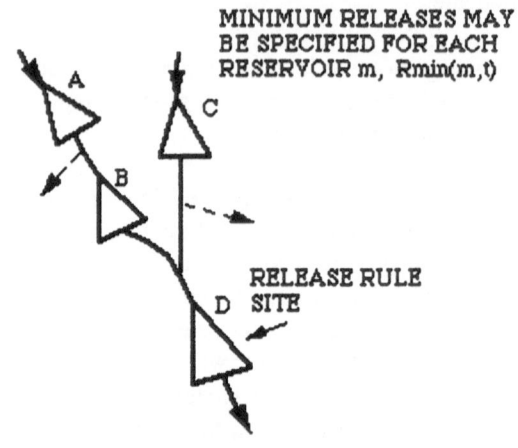

MINIMUM RELEASES MAY
BE SPECIFIED FOR EACH
RESERVOIR m, Rmin(m,t)

RELEASE RULE
SITE

Total Storage Volume	Target Storage Volume in Reservoir			
	A	B	C	D
10	5	0	5	0
20	10	0	10	0
60	30	5	20	5
100	30	25	25	20
200	30	45	60	65

Figure 5. Multiple reservoir storage targets are expressed as functions of the total storage volume in all reservoirs of the group. These functions are defined for all reservoirs not at reservoir release rule sites. The target storage volumes are defined by linear interpolations between the volumes listed in the table. They can vary in each within-year period.

9. Simulation Procedure

The IRAS simulation takes place in a separate program module. The simulation takes place in daily time steps or in time step durations of at least one-twelfth of the within-year period, whichever is shorter. Considerably shorter than daily simulation time steps may be required if water quality constituent concentrations are being simulated.

The simulation procedure involves a number of passes or loops through various system components and data sets. These include the hydrologic flow sequences, the various time periods being considered in the simulation, and the network nodes and links. As illustrated in Figure 6, the simulation begins by first identifying a hydrologic sequence. For each of the possible multiple hydrologic sequences (or replicates), the simulation proceeds

SIMULATION PROCEDURE

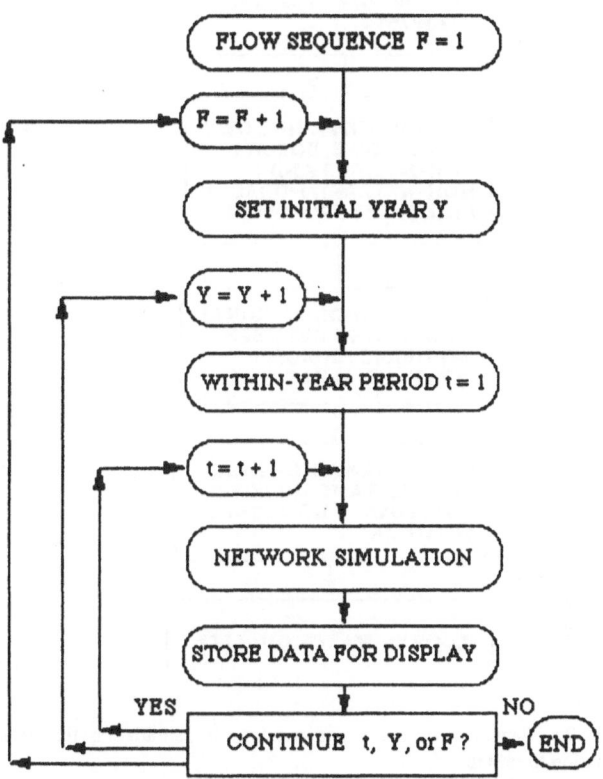

Figure 6. Overall simulation structure used in IRAS.

through the sequence of successive within-year time periods, from one year to the next, through all years of the simulation. Within each within-year period, the simulation proceeds from one simulation time step to another.

184

   In each simulation time step of each within-year period, the simulation proceeds through the system network a number of times to determine the flows, storage volumes, evaporation and seepage losses, and hydropower production and quality at each node and link, as applicable. Figure 7 shows these separate simulation loops through the applicable parts of the node-link network.

NETWORK SIMULATION LOOPS

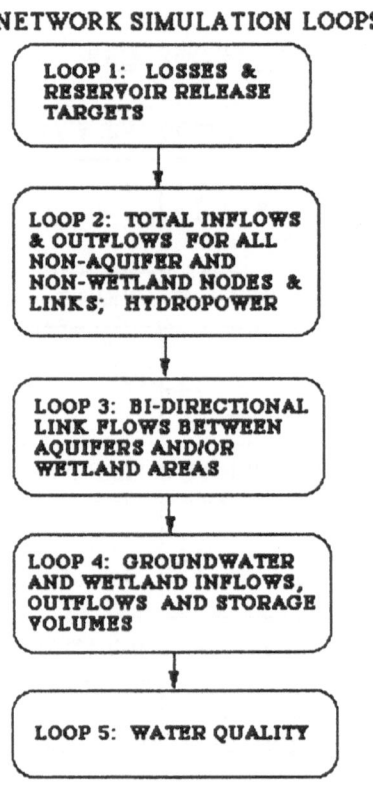

Figure 7. Sequence of operations required for simulating the network in each simulation time step.

10. Simulation Output

   The results of any simulation run are the initial or final storage volume values and average flow, energy, and water quality values for each within-year period, expressed in the units defined by the user. These data can be plotted over time or space. Space plots can be dynamic; thus showing how some variable value changes over space and time. User-defined functions of these output variables as well as statistical analyses based on these output variables can also be computed and displayed. These displays can include probability distributions of resilience and vulnerability criteria, the latter based on either duration of failure and extent of failure. Once created, the output files can be used for the direct display of the results of the simulation or they can be used by a utility program designed to transfer user-selected data to spreadsheet programs for further analyses and display.

10.1   LINKING HYDROLOGY TO ECOLOGY

Of interest to many involved in watershed and river basin management are not only the hydrologic and environmental impacts of various land and water management policies, but the corresponding ecological consequences. The model outlined below attempts to define the health or 'integrity' (Karr 1991) of stream corridor systems that includes the riparian zone using functions of hydrosystem indicators and the time-series outputs of simulations. The procedures require the definition of the hydrosystem integrity indicators, functions of habitat stability and water quality, and the time-series of simulated. Once the integrity indicators are defined, ecologists must be able to relate indicator values to the quality of health of stream corridor systems. This is done through the definition of suitability functions for each indicator. In our model, the indicators and suitability functions have been defined mostly from established ecological principles and realtions, and in some aspects by the judgement of ecologists in the region.

10.2   HYDROSYSTEM   INTEGRITY   INDICATORS   AND SUITABILITY FUNCTIONS

Our indicators can be grouped into two classes:

- Indicators that capture the variability of the habitat or structure of the stream environment, and
- Indicators of the chemical and physical state of the stream system environment that will vary over time, say from day to day, but are related to hydrosystem integrity by the values attained through time.

We define the former as habitat indicators and the latter as water quality indicators. Both vary over space, i.e. from site to site along a stream or river within a watershed or river basin. They differ in the way the indicators are defined and how the suitability functions are expressed.

10.3   HABITAT INTEGRITY INDICATORS AND SUITABILITY FUNCTIONS

Five integrity indicators (Figure 8) are used in our model to capture the suitability of stream habitat to support a diverse and abundant aquatic biota. We defined the indicators based on the practical consideration of what measures could be developed in our hydrology simulations, and how well we could relate feasible measures to the biological capacity of stream habitat. These five indicators are not likely to be comprehensive of all model needs in all cases and geographic settings. For example, stream contiguity is not an important factor in our, but if it were the extent of contiguity could also be given suitability values.

The first habitat integrity indicator (Figure 8), flow persistence, relates the predictability of streamflow occurrence to habitat suitability. As the existence of streamflow becomes unreliable, the potential support for stream organisms and riparian vegetation declines (Poff et al. 1997). While we found little quantitative information to parameterize this indicator, we expect that streams averaging more than 90 days a year without flow will offer no significant habitat for flowing water biota.

Flow stability is an indicator of the variability of stream volume (discharge) defined as the ratio of the range of the 10%ile to 90%ile discharges (annual or longer time series) adjusted by the mean discharge (Figure 8). Very large values indicate elevated flow variation generally caused by runoff from farmland and highly developed land in our watersheds. Elevated flow variation is associated with degraded stream communities (Poff and Ward 1989), and we based the suitability function on judgement.

As with elevated flow variability, we expect that increased thermal fluctuations would be associated with degraded habitat suitability. Reduced base streamflow and tree shading would contribute to greater variation in water temperature with changing weather conditions. Our temperature stability (Figure 8) indicator is our initial estimate of this suitability function.

The substrate stability indicator, defined as the average daily percent streambed surface replacement, reflects reduced habitat quality when stream bottom material is frequently in motion. Mobile streambeds are usually composed of fine material that provides poor substrate for bottom organisms, and when the material moves the attached biota is lost. This function is based on Cobb, et al. (1992) who related abundance, diversity and distribution of benthic insects to substrate stability.

The final habitat indicator captures variability in spatial structural complexity rather than time. Gravel, cobble, and boulder bed materials provide complex surfaces that provide abundant crevices and inter-particle spaces for invertebrates. Also, these coarse materials allow water flow through the surface substrate and biological production within the substrate can be an important component of stream communities. In contrast, fine material like silt and sand provide few living surfaces as does large rock and bedrock (Waters

1995). Our substrate composition indicator (Figure 8) captures this habitat suitability relation.

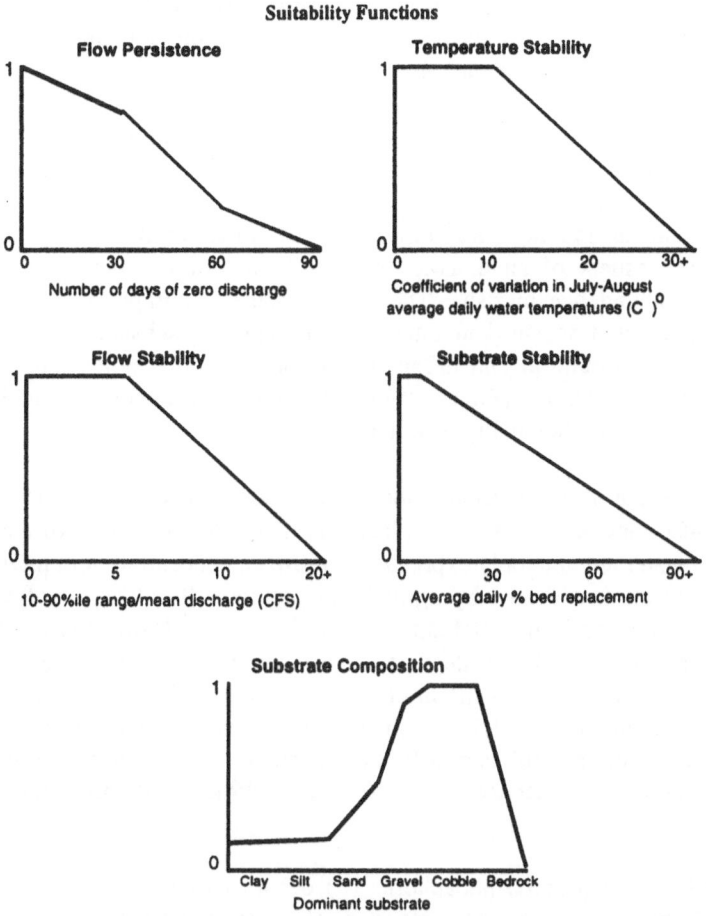

Figure 8. Five habitat integrity indicator suitability functions emphasizing the stability and spatial complexity of stream environments for the support of a diverse and abundant biota.

10.4   HABITAT INTEGRITY VALUE

For a specific site, suitability values, S_i, for each of these habitat indicators, i, can be obtained. Here there are 5 such values. In general there could be i = 1,2,...,n such values. Assuming all indicators are equally important, the geometric mean habitat suitability value, S_H, can be computed as the nth root of the product of the n suitability values, S_i.

$$S_H = [\, \Pi_i^n \, S_i \,]^{1/n}$$

The geometric mean (which will generally be a lower value than the arithmetic mean) is proposed for use as an overall indicator of habitat integrity for any site. The geometric mean will be 0 if any one of the indicators is 0. This seems reasonable. By definition, an indicator value of 0 will lead to death, i.e., an the hydrosystem cannot support a intact biota under those conditions no matter how suitable the other indicator values.

10.5 WATER QUALITY INDICATORS AND SUITABILITY FUNCTIONS

The chemical and material composition of water, or its quality, is often the primary aspect of an aquatic habitat that determines the species composition and abundance of fish. The leading causes of impairment are excessive levels of siltation, nutrients, and oxygen depleting substances most often the result of agricultural and urban runoff and municipal point source pollution. Our four indicators (Figure 9) include the most common chemical parameters used to assess the quality of water.

Nitrate is one of two nutrients (the other, phosphorus described next) important because they are required by plants and algae for growth. Nitrate levels in stream water are generally expressed as nitrogen concentration in mg/l (Figure 9). Nitrate is elevated by agricultural pollution (fertilizer runoff), sewage (breakdown of proteins), and atmospheric deposition. Artificially high nitrate concentrations in surface waters pose the potential for contributing to eutrophication and promoting a shift in community composition to pollution tolerant species. Smith et al. (1993) use a nitrate concentration of 1 mg/l N as indicative of agricultural and urban runoff effects, and 10 mg/l is considered by the U. S. Environmental Protection Agency (US EPA 1986) as the maximum quality limit.

In surface freshwaters, phosphorus (P) occurs largely as phosphate (PO4) and phosphorus is often the component of water chemistry that limits plant production. In natural stream settings, phosphorus rapidly taken up by living things. However, many human activities greatly increase the availability of phosphorus as phosphate resulting in accelerated growth of alga, eutrophication, and community composition shifts. The sources of phosphate to streams are generally waste discharges from point-sources and nonpoint sources of phosphorus include agricultural and urban runoff. No serious problems usually occur if phosphate concentrations remain under 0.1 ppm or mg/l and the U. S. Environmental Protection Agency (EPA 1986) uses this as the quality criteria definition for flowing waters. For standing waters, a maximum quality limit is 0.5 mg/l is often used.

Suitability Functions

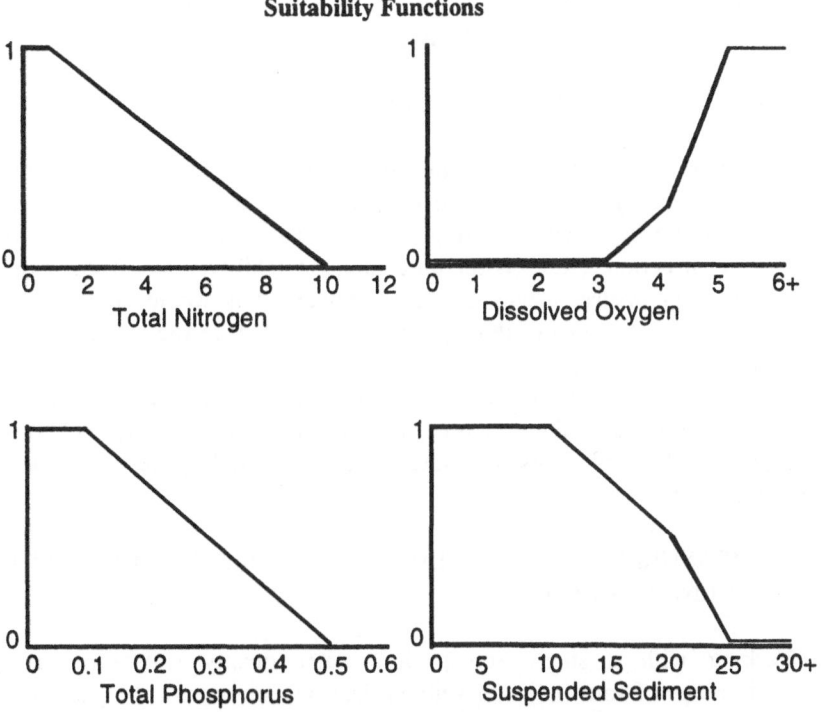

Figure 9. Four water quality indicator suitability functions based on common chemical and material concentrations in relation to the support of a diverse and abundant biota.

Adequate dissolved oxygen is essential for aquatic biota living in the water and sediments. For these reasons, dissolved oxygen has historically been one of the most frequently measured indicators of water quality. The U. S. Environmental Protection Agency (US EPA 1986) specifies a minimum concentration of dissolved oxygen of 5 mg/l as a quality criteria for maintenance of aquatic biota, and some states use a less protective minimum limit of 4 mg/l. These values were used to specify our suitability function as seen in Figure 9.

Suspended sediments are fine particles, primarily clays, silts, and fine sands, that require only low velocities and minor turbulence to remain in suspension. A variety of factors may increase the sediment load in streams to a level that is detrimental to their biological communities (Waters 1995). Excessive sedimentation usually occurs as the result of human activity, including but not limited to agriculture, logging, mining, and urban development. Suspended sediment concentrations vary greatly in natural settings due to local geology and soil types, and clear criteria for effects on aquatic biota are not available. As an initial guide for our water quality indicator, we use 10 mg/l as the upper limit of fully suitable water quality for our region (Figure 9). About 25% of US streams have average suspended sediment levels less than 18 mg/l (Smith et al. 1987).

10.6   WATER QUALITY INTEGRITY VALUES

Single measurements or simulated values of most water quality parameters are highly variable due to temperature, season, stream discharge, and recent watershed activities (e.g, forest fires, land clearing, farming practices, construction, etc.). While an average suitability value of the entire time series of geometric means can be computed, this would shroud time periods of poor water quality thought to heavily influence the aquatic biota. Therefore, we computer a single geometric mean of the four suitability values for any day or time as shown above for the habitat indicators. Then, the geometric mean is assessed over a time series of mean water quality suitability values to develop measures of water quality reliability, resilience and vulnerability.

10.7   MEASURES   OF   RELIABILITY,   RESILIENCE   AND VULNERABILITY

Working with a single (composite) water quality suitability value (geometric mean of four indicator suitabilities), we summarize the pattern of suitability variation to make a summary assessment. The composite water quality suitability value (0,1) is viewed as being within one of three zones :
- a high quality (green) zone,
- a marginal quality (yellow) zone, and
- an degraded quality (red) zone.

Once these three suitability zones are defined for one or more selected geographic sites, various time series statistics can be computed and displayed at each of those selected sites in the watershed. The computed statistics include:
- green and yellow zone reliabilities (the probabilities of being in each of those zones in any simulation time period).
- yellow and red zone resiliences (the probabilities of being in the green or in the yellow or green zones following a value in the yellow or red zones, respectively, in the next simulation time step).
- yellow and red zone vulnerability probabilities based on the extent of 'failure' in an individual (single) time period and/or the sum of extents over a sequence of successive time periods
- yellow and red zone vulnerability probabilities based on the duration of consecutive sequence of suitability values in those 'failure' zones.

Figure 10 illustrates a sample time series of mean suitability values, and the three zones of suitability values as defined by the user. These zones may be time dependent, i.e., they could vary within each year. However defined, they are the basis for the computation of the reliability, resilience and vulnerability statistics. The example in Figure 10 includes 42 time periods. Only 8 of those periods have mean suitability values in the green (satisfactory) zone and 14 periods have values in the red (unsatisfactory) zone. The remaining 20 periods have values in the yellow (warning) zone. From this we can calculate the green reliability is $8/42 = 0.24$. Yellow reliability is $(8+20)/42 = 0.67$. Only once in this illustrative sample did a 'yellow' value become a green value (barely) in the following period, hence yellow resilience is $1/20 = 0.05$. Red zone values became yellow zone (or green zone) values 3 out of 14 opportunities, hence red zone resilience is $3/14 = 0.2$.

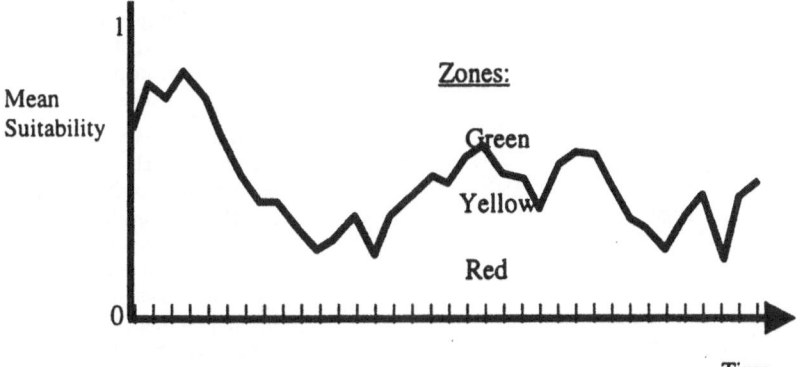

Figure 10. Time series of mean water quality suitability values resulting from a simulation of a land and water management scenario. Three suitability zones are shown.

Vulnerability measures can be expressed as the probability of

- an individual extent of failure (the difference between the minimum green or yellow zone value and the actual value in the yellow or red zone respectively), or
- a cumulative extent of failure (the difference between the minimum green or yellow zone value and the sum of the consecutive actual values in the yellow or red zone respectively), or
- a duration of failure (number of time periods in which suitability values are in the yellow and red, or red zones).

192

Table 1. Interpretation of stream integrity based on simulated habitat and water quality indicators.

Habitat quality	Water quality reliability (P[green])	Water quality resilience (P[red->non-red])	Water quality vulnerability (P[red])	Interpretation
High	High	High	Very low	A high integrity stream with intact habitat and consistently good water quality.
Low	High	High	Very Low	A physically altered stream with good water quality. Common situation for channelization and hydropower development in sparsely populated regions.
High	Low	Low	High	A degraded water quality stream with intact habitat. Most likely the result of intense land uses that span uplands of a watershed. Near stream and riparian areas are probably in natural vegetation.
High	High	Low	Moderate	A stream with intact habitat and mostly good water quality that occasionally experiences stressful water quality for extended periods of time. This condition could result from seasonal agricultural practices that impact water quality strongly for several weeks each year.
Low	Moderate	High	Moderate	A stream with degraded habitat and highly variable water quality that changes quickly. This condition would likely be seen in streams draining suburban landscapes where the habitat is mostly channelized and responsive to an array of runoff and land use effects.

We will be using field data together with these measures of water quality suitability together with the overall habitat suitability value, S_H, to predict ecosystem quality as a function of land and water management policies and practices.

10.8   FINAL INTERPRETATION OF HYDROSYSTEM INTEGRITY

Our set of nine suitability indicators are combined into two integrity values: habitat and water quality. The habitat indicators use long time series of values to assess suitability as a single indicator value; e.g. flow regime is assessed as one flow stability value. Habitat was modeled as a fixed property under any single land use and environmental setting because habitat quality is a product of long-term processes. In contrast, water quality changes rapidly in response to many, often short-term human activities like construction, farming practices, and climatic events. For this dimension of integrity, we assess the mean water quality suitability using measures of a time series of values (as in Figure 10). The summary conclusions about the health or integrity of the hydrosystem requires a synthesis of habitat and water quality indicators. We propose to make our final interpretation without a quantitative integration of the two final indicator results (Table 1).

194

References

Bain, M. B., J. T. Finn, and H. E. Booke. 1988. Streamflow regulation and fish community structure. Ecology 69:382-392.

Cobb, D.G, Galloway, T.D., and Flannagan, J.F., 1992, Effects of discharge and substrate stability on density and species composition of stream insects, Can. J. Fish., Aquat. Sci. 49: 1788-1795.

Karr, J. R. 1991. Biological integrity: a long-neglected aspect of water resource management. Ecological Applications 1:66-84.

Poff, N. L. and J. V. Ward. 1989. Implications of streamflow variability and predictability for lotic community structure: a regional analysis of streamflow patterns. Canadian Journal of Fisheries and Aquatic Sciences 46:1805-1818.

Poff, N. L., J. D. Allan, M. B. Bain, J. R. Karr, K. L. Prestegaard, B. D. Richter, R. E. Sparks, and J. C. Stromberg. 1997. The natural flow regime: a paradigm for river conservation and restoration. BioScience 47:769-784.

Smith, R. A., R. B. Alexander, and M. G. Wolman. 1987. Water-quality trends in the Nation's rivers. Science 235:1607-1615.

Smith, R. A., R. B. Alexander, and K. J. Lanfear. 1993. Stream water quality in the conterminous United States -- status and trends of selected indicators during the 1980's. U.S. Geological Survey Water Supply Paper 2400.

U. S. Environmental Protection Agency. 1986. Quality criteria for water, 1986. U. S. Environmental Protection Agency, Office of Water Regulations and Standards, EPA 440/5-86-001, Washington, DC.

Waters, T.F. 1995. Sediment in Streams. American Fisheries Society Monograph 7, Bethesda, Maryland, USA.

A DECISION-MAKING SUPPORT SYSTEM FOR SELECTING WATER PROTECTION AND RESERVOIR CONTROL MEASURES

V.G. PRIAZHINSKAYA
Water Problems Institute, Russian Academy of Sciences
Ul. Gubkina, 3, Moscow 117971, Russia.
L.K. LEVIT-GOUREVICH
Institute of Water System Design (VODNIIINFORMPROEKT)
Ul. Kedrova, 14, bld. 2, Moscow 117078 Moscow, Russia
D.M. YAROSHEVSKII
Water Problems Institute, Russian Academy of Sciences
Ul. Gubkina, 3, Moscow 117971, Russia

1. Introduction

The development of computer technology over the last years allowed creation of integrated decision-making support systems (DMSS) for the water economy. Such systems reflect the decision-maker's (DM) desire to consider a water object as a whole with all technical, ecological, financial, economic and other problems related to it. They provide tools for analyses of different solutions allowing to take into account all formal and informal considerations involved in planning, design and management of water resource system (WRS). DMSS offers a common platform for solving complex water problems utilizing *decomposition* principles.

The suggested methodology is suitable for solving the majority of water problems in a large river basin, where problems of different nature often arise. That is why it is necessary not only to make a systems analysis (decomposition of the problems, engineering statements, mathematical models, and computer realization) but also to investigate the methods for co-ordination of separate tasks and problems to obtain a complex solution of the water problem. In this paper a methodology for solving two water problems is briefly described, illustrating an attempt of a search for complex solutions.

2. Methodological Principles of Problem Structuring

A mathematical solution of complicated water problems using only one model is often impossible (and inexpedient). Therefore it is necessary to split the *complex problem* into separate *sub-problems*. However, a choice of the most effective decomposition method is non-trivial. Splitting according to various water disciplines is appropriate, but a certain problem belonging to some field of water sciences cannot be solved completely

M.V. Bolgov et al. (eds.), Hydrological Models for Environmental Management, 195–213.

196

and for the whole large river basins at once. A necessity for solving this problem for its parts thus arises with a consequent coordination of the results for the whole basin framework. The aim of the decomposition is ensuring the homogeneity in the complexity and degree of details in the mathematical models of all DMSS elements. This is achieved with the help of a co-ordination between the degree of details in the tasks' statements and the scale of water objects concerned [3].

A central element of the DMSS is a *subsystem* defined as an aggregate of a certain problem and a certain part of a river basin. Subsystems belonging to the same water discipline for all territorial and river *units* (parts of the basin) build a *problem block*. Each subsystem includes one or several tasks. A task is defined as a part of the problem (with its engineering statement and mathematical formalization) that can be solved separately. Before describing subsystems and tasks in each problem block it is necessary to subdivide the studied river basin into units based on administrative, geographic, hydrographic and other important criteria. This partitioning is based on the analysis of homogeneity and uniformity of the objects. Another demand to the partitioning is hierarchy of the units that allows *aggregation* and *disaggregation* of the results between subsystems or tasks of the next hierarchical levels.

The links between the tasks of one subsystem or between the subsystems with different problems are based on the information exchange of the type *"input-output"*. All types of the information exchange allow realizing an iterative search for the global solution in the whole DMSS framework. A formalization of these iterative procedures for many problems and subsystems represent the global mathematical decision model *(meta-model)* [3]. Each subsystem includes usually: *optimization* tasks for a choice of the main parameters and characteristics of an object, *simulation* tasks for determining probability characteristics for object's functioning, tasks of *unitary modelling* for obtaining technical parameters and characteristics of an object.

In a DMSS for a large river basin some problem blocks are distinguished according to the disciplinary decomposition and grouped into *natural process modelling, engineering systems modelling, water economy modelling, water objects' analysis,* and *water organization control*. Each group may be also subdivided into various separate problems. For example, the *engineering systems modelling* group in the Volga River basin is divided further into a number of subproblems, of which the two most interesting here are *water quality management* and *water resources control*. Also the combinations of water resources and economic objects are considered by means of a decomposition of a large basin into administrative units (provinces, districts etc.) and hydrographical units (rivers or river sections of a various length).

3. Water Quality Management Tasks

An example of a DMSS problem block is *water quality management*. It consists of six subsystems: the upper level of large water bodies like provinces and large river sections;

the middle level like districts and river sections of the intermediate scale; the lower level including small rivers and localities. Territories and river sections represent two independent systems at these three levels. All territories and river sections at each level are linked through water entering conditions. A river section with a certain territory attached to it is called *the main river (river section)* for this territory. According to the partitioning principles the main rivers for territories at one level may be included into objects of other levels. Thus, at each level there are two subsystems *viz.* territorial and basin; both including optimization and simulation tasks.

Optimization tasks <u>for territories</u> involve a selection of the best complex of protection measures at low costs ensuring normal water quality in water objects on the territories and leading to a limited amount of polluted water to the main river section within each territory. Different measures may be chosen depending on feasible emission of pollutants. Optimization tasks <u>for a river basin</u> involve a choice of a general solution for protection measures in the whole basin at low costs for normal water quality conditions in a river. <u>Simulation tasks</u> for territories and rivers involve estimation of the probability characteristics for functioning of protection measures. Subsystems of the upper, middle and lower levels differ in details of representation of the protection measures and water quality.

All protection measures are subdivided into the following *classes*: treatment of point source pollution effluents, treatment of cattle-breeding farms, runoff management in urban areas, organization of drainage from agricultural lands, river water protection zones, protective measures in river basins and reservoirs. Each class of protection measures is divided into various *groups*. For example, measures in a river basin include protection against pollution by water transport, and protections against pollution by rafting and water aeration. Each group consists of several *types* of specific protection measures.

Water quality is characterized by quality classes in a general description (according to existing standards), by concentrations of the main pollution groups or by concentrations of certain pollutants in detailed descriptions. The subsystems of the upper level operate with classes of protection measures and quality classes. Subsystems of the middle level operate with groups of protection measures and concentrations of the main pollution groups. The subsystems of the lower level operate with particular types of protection measures and pollutants.

The simulation tasks examine water quality characteristics as a result of protection measures during a long-term period. The runoff and river discharges in these tasks are random variables and vary in time but pollutant loads are constant. Transformation of pollutants in the river is evaluated by simulation tasks. These evaluations are carried out by means of a simple mathematical model, which describes pollutant decomposition by concentration functions C :

$$C = C_0 \exp(-kt) = C_0 \exp(-kL/v), \quad C_0 = m_0 /(Q+q) \tag{1}$$

where: C_0 - primary concentration in the inlet of a river section, m_0 - emission mass, Q - river flow, q - wastewater flow, k - decay rate, t - travel time, L - length of a river section, v - average flow velocity. This equation based on the principle decomposition of the first degree was suggested in 1920 by Streeter and Phelps.

A generalized mathematical formulation of the least-cost management models will be presented for two tasks involving a choice of protection measures *viz.* territorial and basin tasks. The following symbols are introduced: m - number of the protection measure, i - number of the territory, r - sections, g - number of the pollutant, I_r - set of territories for which r is the main river section, R_r - set of river sections preceding section r. Furthermore S_{mi} denotes the cost of measure m defined as a function of annual operation and maintenance costs and amortization costs; p_{mi} are parameters of the m-th measure; G^+_{gmi} and G^-_{gmi} - pollutant loads of the g-th pollutant before and after implementation of the protection measure; G_{gi} are the total pollutant loads on the territory; e_{gi} and e_{gr} - coefficients ($e < 1$) for distributing the pollutant loads among water objects on the territory and the main river section; w_i is the volume of water resources for a given time period; and d_{gi} are the coefficients characterizing decomposition of pollutants in local water.

For the first task it is necessary to find the complex of protection measures for the territory corresponding to the minimum total cost, satisfying the demands for water quality in water objects and limiting pollutant loads G_{gir} entering the main river from this territory. The necessary model can be formulated as

$$S_i = \sum_m S_{mi}(p_{mi}) \to \mathbf{min} \ . \tag{2}$$

The pollutant loads reduce as a result of protection measures. They can be defined by the *functions of the efficiency of the protection measures*:

$$G^-_{gmi} = E_{gm}(p_{mi}, G^+_{gmi}) \text{ for all actions } . \tag{3}$$

The distribution of pollution mass among the local water objects on the territory and in the main river section is determined as

$$G_{gi} e_{gi} = \sum_{m \in M_i} G^+_{gmi} \tag{4}$$

$$G_{gi}e_{gr} = \sum_{m \in M_r} G_{gmi}^+ \quad . \tag{5}$$

The limiting concentrations C_{gi} for a local water object of the normative character $\overline{C_g}$ are obtained as

$$C_{gi} = \left(\sum_{m \in M_i} G_{gmi}^- \right) / w_i \leq \overline{C_g} \quad . \tag{6}$$

The limit for the pollution volume that is allowed to enter the main river is found as:

$$d_{gi} \cdot \sum_{m \in M_i} G_{gmi}^- + \sum_{m \in M_r} G_{gmi}^- \leq G_{gir} \quad . \tag{7}$$

Performing this formalization with an economic criterion the minimum costs are controlling the choice of the measures satisfying water quality demands. With an ecological criterion the maximum efficiency of the protection measures it is necessary to reach the normative conditions defined by eqs. (3)-(5) at a limited cost. The decision - maker may choose any modification of the task and use any results. In principle several complexes of measures are determined by the first task depending on the predefined G_{gir} , i.e. feasible pollutant loads in the main river coming from the territory after the protection measures have been carried out.

In the second task it necessary to select protection measures for each territory adjacent to the main river and also in the river basin in such a way that this ensemble of measures would allow satisfying the demands to water quality at minimum cost, i.e:

$$S = \sum_r \left(\sum_{i \in I_r} S_i + \sum_{m \in M} S_{mr}(p_{mr}) \right) \to \mathbf{min} \quad . \tag{8}$$

Assuming that the pollutant loads obtained after the protection measures have been carried out are described by *the function of the efficiency of the protection measures*

$$G_{gmr}^- = E_{gm}\left(p_{mr}, G_{gmr}^+\right) \text{ for all that measures} \tag{9}$$

with the pollutant loads defined as

$$G_{gr} = \sum_{m \in M} G_{gmr}^+ \tag{10}$$

and concentration limits of the normative character $\overline{C_g}$ defined as

$$\left(\sum_{i \in I_r} G_{gir} + \sum_{\rho \in R_r} (G_{gi\rho} \cdot d_{g\rho}) + \sum_{m \in M} G_{gmr}^- \right) / W_r \leq \overline{C_g} \tag{11}$$

where $S_{mr}, p_{mr}, G_{gmr}^+, G_{gmr}^-$ are respectively the cost, the parameters of the protection measures in the river basin and pollutant loads before and after the measures have been carried out, W_r are water volumes in the cross-section for a certain time period; and d_{gr} is the decay rate of a chemical compound g.

A topological tree pattern of a river network and a functional independence of the protection measures usually allow a stepwise treatment of the conditions and limitations and additivity of the objective function. Thus, a utilization of a stepwise scheme of dynamic programming is possible. In certain cases of aggregation the objective function, conditions and limitations can be created using linearization with an employment of standard methods of linear programming. A common linear optimization model of a regional wastewater treatment policy can be formulated as

$$\left\{\min S\varepsilon \,\middle|\, A\varepsilon \leq B, 0 \leq \varepsilon \leq 1\right\} \tag{12}$$

where S is the treatment cost (total annual cost - TAC- or total investment cost - TIC), A is a matrix of technical and economic coefficients, which includes a feasible set of treatment alternatives and their effects on water quality in a river, B is a vector of resource limitations, and ε is a vector of the proportions of different treatment technologies. All water quality characteristics in a river, the efficiency of the protection measures, the volumes of water resources, and the processes of pollutant transformation can be assessed in a linear form. The simulation model contained in the optimization procedures estimates the consequences of a variety of feasible alternatives. It is necessary to note that a decision on the choice of the protection measures (territorial task) depends on feasible pollutant loads G_{gir} entering the main river from the basin. A parametric dependence of the decisions in tasks (2)-(7) on G_{gir} can be represented as *a production function* of water quality management for a territory.

All the tasks for the subsystems are used for planning the water quality management in a large river basin. Information exchange between the territorial and basin tasks at each level and between the tasks at the adjacent levels is possible. The production functions of water quality management for the territories go over from the territorial tasks to the basin tasks; backwards feasible pollutant loads G_{gir} pass from watershed to basin. The solutions chosen within the optimization tasks go over to the simulation tasks. This information exchange corresponds to the "input-output" connection type.

The values for the costs and protection measures efficiency go over from the tasks at the lower (or middle) level to the tasks at the middle (or the higher) level. These values are aggregated from the types (or groups) of the costs and efficiency of the protection measures to the groups (or classes) of the costs and efficiency of the protection measures. Feedback of some information is envisaged to allow diminishing the ensemble of decisions at the lower level tasks. This information exchange corresponds to the type "aggregation-disaggregation" connection.

4. Reservoir Control Tasks

Reservoir control tasks are solved by DMSS treating them separately from the tasks of river basin management. The control subproblem can be further subdivided into sub-problems for different reservoir volumes as in a large river basin it is difficult to consider simultaneously all the reservoirs. Reservoir control tasks for the reservoirs having little effect on the main river flow can be neglected when dealing with the main reservoirs. However, the same tasks can be used for all reservoir systems with different volumes. The formulations and engineering characteristics of all the tasks in the sub-problem can be used for a reservoirs system and an individual reservoir of any size. The following basic tasks in the subproblem can be distinguished:
1. Establishing the dispatch rules for the high water control and lowered reservoir levels during the periods preceding high water conditions.
2. Establishing the dispatch rules for annual and long-term reservoir control.
3. Simulating long-term reservoir operation.
4. Modelling flood runoff for reservoir systems.
5. Simulating water quality in streams and reservoirs.

All the tasks use a common database and common calculation methods. Some examples of the tasks are a preliminary evaluation of water supply and water use for the whole river system or identification of the areas with a risk for flooding by special *rating coefficients*. *A conditional profit* from water is calculated on the basis of these rating coefficients. Establishment of the *dispatch rules for high water control* is an optimization task for minimizing the damages or, on the opposite, maximizing the conditional profit from the areas with a risk for flooding (in case they are not flooded) and high reservoir levels during periods following the floods. The task is carried out basing on a set of hydrograph schemes.

The establishment of the *dispatch rules for reservoir control* is an optimization task for maximizing the conditional profit from water supply and water use. The dispatch rules for reservoir control are described by the mathematical functions for a control time period and reservoir. The parameters of these functions are calculated as a part of the optimization tasks.

The simulation task involves a determination of the probability characteristics of water supply, water use, areas with a risk for flooding, i.e. the quality of the established dispatch rules is verified.

Modelling flood runoff involves (as the only operation) a determination of the precise reservoir characteristics from the observed flood hydrograph.

Simulation of water quality in a river system with reservoirs involves a determination of pollutant concentrations in all river cross-sections for long-term time periods with fixed pollutant loads. The results are necessary for assessing permission for pollutant emissions. The connections between all these tasks are based on the information exchange of "input-output" type. This allows adjusting the dispatch rules in all contradictory situations.

A complete formalization and mathematical models for the choice of dispatch rules for reservoir control and problem solution methods are described in [2]. Here the generalized mathematical models and principles they are based on are discussed. The following symbols are used: r - number of a reservoir/river section; t - control time intervals; j - consumers of water supply; k – water users; l - number of a flood area. Expert rating coefficients for the time intervals are denoted as $P_{jt}, R_{kt}, X_{lt}, Y_{rt}$. It is assumed that all these coefficients are normalized such that: $\sum_j P_{jt} + \sum_k R_{kt} = \sum_l X_{lt} + \sum_r Y_{rt} = 1$.

The necessary *condition for water supply* is a supply of the predefined (normative) amounts of water to all the consumers during each individual year of a long-term period with a probability not lower than the predefined (normative) for water supply. The necessary *condition for water use* is maintenance of the predefined (normative) levels, depths and water flows necessary for normal operation of the users for a long-term period with a probability not less than the predefined (normative) for water use. For *a hydropower plant*, the minimum permissible power production is also fixed, and this value is connected to the product $Q \cdot H$, which is a tail-water discharge of the power plant multiplied by the water head. The necessary *conditions for flood water passage* consist in a restriction of the water depth on the flooded areas or of the flood levels for a long-term period with a probability not less than the predefined (normative) probability of the maximum flood levels. All these conditions can be written as follows (in the order of their presentation):

$$Q_{jt}^p \leq \overline{Q_{jt}^p} \tag{13}$$

$$\underline{Z_{rt}} \leq Z_{rt} \leq \overline{Z_{rt}} \tag{14}$$

$$(QH)_r \leq Q_{rt} H_{rt} \tag{15}$$

$$\underline{H_{lt}} \leq H_{lt} \leq \overline{H_{lt}} \; . \tag{16}$$

Since the demands of water consumers, users, and land protection from flooding are contradictory in many cases, the normative probabilities can be difficult (or impossible) to establish. In such cases, if the dispatch control parameters are to be determined on the basis of water economy calculations, the normative values should be corrected. Calculations with respect to water economy are used to analyze the possibility of satisfying all the above conditions and to correct the relevant water economy parameters.

A *conditional profit* $D_t = D_t^r + D_t^h$ from a water economic system is calculated on the basis of the rating coefficients for water supply consumers, water users and lands with a risk for flooding determined for the t-th interval, where D_t^r is a conditional profit obtained from water supply customers and water users by means of water resources control, and D_t^h is a conditional profit obtained from the lands with a risk for flooding (if they are not affected by a flood) by means of flood control.

A conditional profit with fixed normative water discharges and reservoir levels depend on the reservoir characteristics for a time interval, such as: water tail discharge of a reservoir sluice-regulator Q_r; total flow volume for water supply in a river section Q_r^p; reservoir level Z_r (indirectly depending on preceding values). Thus, the conditional profit as a function of the reservoir state parameters has the following form:

$$D_t(Q_n, Q_n^p, Z_n) = D_t^r(Q_n, Q_n^p, Z_n) + D_t^h(Q_n, Z_n) \ . \qquad (17)$$

The *dispatch rules for reservoir control* establish *the control parameters* based on the *parameters of a reservoir state*. The main parameters of the reservoir state are calendar date T and reservoir level Z_{it}. Forecasts of flood characteristics can be added to these in the case of flood control. In the case of regulation of low flow, filling up the upstream and downstream reservoirs can be added to each reservoir state parameters. The following relationships describe the dispatch rules for flood control U_h and for water resources control U_r:

$$Q_n = U_h(t, Z_n); \quad (Q_n^p, Q_n) = U_r(t, Z_n) \ . \qquad (18)$$

The rules assure a guaranteed (minimum permissible) reservoir tail discharge Q_n^g. A possibility of increasing the reservoir tail discharge to $Q_n > Q_n^g$ appears in case of additional water inflow to the reservoir during the time interval considered.

Before discussing mathematical models for reservoir control it is useful to formulate in advance some technical and physical conditions, which concern all the models to a certain extent.

The condition of water balance for a design time control interval is formulated as:

$$\Delta V / \Delta T = Q^+ - Q^- \qquad (19)$$

where Q^+ is a sum of all water balance inflow components (runoff from the river basin, outflow and infiltration from upstream reservoirs, etc.), Q^- is a sum of all outflow components (water supply, losses for infiltration, evaporation and ice-formation, reservoir tail discharge etc.), and ΔV is a change of a reservoir volume during time interval ΔT. Limitation for level changes in reservoirs can be written as:

$$Z_r^{dead} \leq Z_{rt} \leq Z_r^{fors} \quad \textit{for high water periods};$$
$$Z_r^{dead} \leq Z_{rt} \leq Z_r^{full} \quad \textit{for low water periods}. \tag{20}$$

The condition of tail-water discharge through reservoir locks is written as

$$Q_r = F_r\left(Z_r^{up}, Z_r^{down}\right). \tag{21}$$

In eq.(20) the dead and banked (normal) water levels in a reservoir are given, while the upstream and downstream water levels of reservoir lock are used in eq. (21).

Establishing the dispatch rules for high-water passage represent an optimization task of maximizing the conditional profit from areas with a risk for flooding and reservoir filling up:

$$D = D_t^h + \sum_r \left(Y_r \cdot V_r(T_1)\right) \rightarrow \mathbf{max}. \tag{22}$$

For the conditions in eqs. (17)-(21), the initial condition is

$$Z_r(T_0) = Z_r^{lower} \tag{23}$$

and the final condition is

$$Q_r(T_1) = U_r\left(T_1, Z_n\right) \tag{24}$$

where T_0, T_1 - beginning and end of a high water period, and Z_r^{lower} – lowered reservoir levels during the period preceding the flood. The sum in eq. (22) gives preference to the alternatives with filling up the reservoirs at the end of the high water period. The condition in eq. (24) links the first and the second tasks.

A mathematical formalization for establishing the dispatch rules for the control of the water resources in a reservoir during a period of one year has the following form: for every control period t necessary to define $Q_{rt}(Z_n)$ and $Q_{rt}^p(Z_n)$ for all reservoirs r that provide

$$D = \sum_t \sum_r D_t^r \rightarrow \mathbf{max} \tag{25}$$

for the conditions eqs. (13)-(21). Here the conditional profit from the water economic system is maximized for a year as the sum of profits for reservoirs and all control periods.

In the task of dispatch rules for the control of water resources in a reservoir an optimization is carried out for one year or a couple of years only and is not intended for long-term periods. The final choice of the dispatch control rules requires examining the solutions of both this and the previous tasks by means of simulations.

The task of simulating long-term reservoir operation ("water resources balance") is carried out by means of simulations of the reservoir functioning for a long-term period based on the fixed time intervals in the adopted dispatch rules. The following characteristics are calculated: tail-water discharges, variations in water levels and volumes, losses for evaporation, infiltration, ice formation, water supply volumes (normative values and reduced values for the years with low water yield), and water use parameters. The calculation results are complemented by a statistical analysis of all the characteristics for the whole long-term period. Besides, an estimation of the number of days with water supply deficit or disturbances in water use is performed continuously. This information is necessary for a correction of the control rules options by the decision-maker.

The task of modelling flood runoff is a modelling effort related to modelling hydraulic and certain water economy parameters for an individual flood in a system of reservoirs. The time intervals used in the flood runoff modelling vary from half an hour to a day depending on the closeness of the flood peaks and spatial and temporal scales of the reservoir system. The hydraulic characteristics of discharge locks gates and characteristics of the natural river sections are considered in detail. The calculation of a flood passage is based on the dispatch rules for flood runoff in a reservoir system.

The final form of the dispatch rules for reservoir control is obtained by means of an iterative solution of all the described tasks. In order to ensure the necessary interconnection the formulation of each task includes the variables assigned by the decision-maker or obtained from the solutions of other tasks (calculated parameters). For example, the parameters necessary for the second task include the lowered level in a reservoir during the period preceding a flood, obtained when solving the first task; and vice versa, the calculation parameters in the first task include the guaranteed tail-water discharges directly after the flood, determined when solving the second task. Such data links between all the tasks allow following an iteration procedure for solving the complex problem of the reservoir control.

5. A Simulation Model for the Operation of a Water Resource System

The model analyzed in this section refers to the last stage of the decision-making using the DMSS. It is used to verify the "precise" management parameters and takes account of the factors neglected in the optimization tasks. If the simulation experiments confirm the correctness of a preliminary selection of all the previously determined parameters and management regimes, the decision-maker can select the final solution. If there are any discrepancies between the "desired" parameters and the ones obtained in the process of simulation, it appears to be necessary to return to the optimization tasks with new strategic parameters. The suggested simulation model differs from similar ones as it combines conventional water balance calculations with evaluating water quality and transformation of both point and non-point constituents. The simulation model consists of three main blocks: the preparatory block, the simulation block itself, and the block for processing the results. The simulation has an aim of obtaining a dynamic pattern of WRS operation within the simulation period $T = [\underline{t}; \overline{t}]$, all the WRS strategic parameters and management rules being fixed. The calculations are done over discrete time steps of an arbitrary duration. For all the dynamic parameters the values averaged over each time step are used.

The WRS structure (position and interrelationship of its elements) is presented in a form of an oriented graph $G = (V, A)$, where V is a set of nodes and A is a set of downstream-oriented arcs. Nodes $v \in V$ are images of reservoirs, stream confluence points, etc. Arcs $a \in A$ are images of sections of transportation of water and constituents (rivers, channels, etc.). Water and constituent flows in arcs depend on the stochastic outcomes of natural conditions $\omega \in \Omega$, and therefore they are stochastic functions of time $q_a^g(t, \omega)$. In order to unify the description of the dependencies, a numbering of g for water and constituent flows is adopted, with index $g = 0$ corresponding to water flows, and $g \neq 0$ corresponding to constituent flows.

Let the index "$+a$" describe parameters in the inlet of arc a, and "$-a$" - parameters in the outlet of the same arc. The ratio $\kappa_a^g(t, \omega) = q_{-a}^g(t, \omega) / q_{+a}^g(t, \omega)$ describes the flow transformation in the arc. Water flows are measured by the volumes of water passing through a cross-section of the stream per unit of time, and constituent flows are measured by masses of constituents passing through the same cross-section per unit of time. The concentrations of constituents (in the inlets or outlets of the arcs) are determined as: $c_{\pm a}^g(t, \omega) = q_{\pm a}^g(t, \omega) / q_{\pm a}^0(t, \omega)$. Lag-time τ_a, i.e. the time it takes the flows to pass from the inlet to the outlet of the arc, is also taken into account. Its value is assumed to be the same for both water and constituent flows.

For non-point sources the flows are described by the curves along the arcs. Let L_a denote the length of the section, presented by arc a, and let flow velocity $\dot{x}_a$ along arc a be constant. Then the flow lag-time will be $\tau_a = L_a / \dot{x}_a$, and the lag-time for the flow to pass from the inlet of the arc to an arbitrary point at a distance $x \le L_a$ from the inlet inside it, will be $\eta(x) = x / \dot{x}_a$. The intensity of constituent's inflow is characterized by the curve $\sigma_a^{\downarrow g}(x,t,\omega)$, and the intensity of its outflow is characterized by the curve $\sigma_a^{\uparrow g}(x,t,\omega)$. The difference $\Delta\sigma_a^g(x,t,\omega) = \sigma_a^{\downarrow g}(x,t,\omega) - \sigma_a^{\uparrow g}(x,t,\omega)$ describes the specific flow (per unit length of arc a in the point, located at the distance x from its inlet), added to or withdrawn from the respective flow, that would have been found in point x if the non-point sources did not exist. The total flow along the arc can be described by the following integral:

$$Q_a^g(x,t,\omega) = \left[Q_a^g(0,t - \eta(x),\omega) + \int_0^x \Delta\sigma_a^g(\xi,t - (x-\xi)/\dot{x}_a,\omega)d\xi \right]^{\oplus} \tag{26}$$

with the initial conditions for the input $Q_a^g(0,t,\omega) = q_{+a}^g(t,\omega)$ (at $x = 0$). In eq. (26) the index $f^{\oplus} = \max(0; f)$ is used for term f in the square brackets. This is explained by the fact that $Q_a^g(x,t,\omega) \ge 0$. Knowing the water volume $Q_a^0(x,t,\omega)$ and the constituent mass $Q_a^g(x,t,\omega)$, the integral concentration values are obtained as $C_a^g(x,t,\omega) = Q_a^g(x,t,\omega)/Q_a^0(x,t,\omega)$.

The simulation is based on balance relations. Let the subset of arcs coming to any node $v \in V$ be denoted as $A_v^+ \subset A$, and the subset of arcs outgoing from the same node as $A_v^- \subset A$; the volume of water (filling) in the v-th reservoir as $W_v^0(t,\omega)$ and the mass of the g-th constituent - as $W_v^g(t,\omega)$. Then, at any instants t_1 and t_2 the law of mass preservation is written as

$$W_v^g(t_2,\omega) - W_v^g(t_1,\omega) = \int_{t_1}^{t_2} \left[\sum_{a \in A_v^+} q_{-a}^g(t,\omega) - \sum_{a \in A_v^-} q_{+a}^g(t,\omega) \right] dt \tag{27}$$

or in the differential form (the continuity equation) as

$$\dot{W}_v^g(t,\omega) = \sum_{a \in A_v^+} q_{-a}^g(t,\omega) - \sum_{a \in A_v^-} q_{+a}^g(t,\omega) . \tag{28}$$

If there is no reservoir in node v, then

$$\sum_{a \in A_v^+} q_{-a}^g(t,\omega) - \sum_{a \in A_v^-} q_{+a}^g(t,\omega) = 0 . \tag{29}$$

The constituent concentration in the reservoir is determined as

208

$$c_v^\delta(t,\omega) = W_v^\delta(t,\omega)/W_v^0(t,\omega) \quad . \tag{30}$$

Note that in the inlets of the arcs outgoing from node v, these concentration values coincide with the constituent concentration values in the source reservoir, i.e.:

$$c_{+a}^\delta(t,\omega) = c_v^\delta(t,\omega) , \quad a \in A_v^- \quad . \tag{31}$$

Therefore, the constituent flows in the inlets of these arcs can be defined using their concentration values as follows:

$$q_{+a}^\delta(t,\omega) = q_{+a}^0(t,\omega)/c_v^\delta(t,\omega) \quad . \tag{32}$$

Eqs. (27) - (32) can be closed by means of assigning the values of initial reservoir fillings and initial masses (or concentrations) of the constituents in these reservoirs (in the beginning of the simulation period).

In order to analyze the results, empirical probabilities of some *critical situations* in the WRS (or in a certain part of it) are used as integral parameters. In this case it is necessary to distinguish clearly the class of events that is called *elementary*; to describe any dynamic parameter as a logical function of these parameters (*composite event*); and to select the form for calculating the respective empirical probability (statistical processing of the composite event).

An elementary event is a result of a comparison of an arbitrary dynamic parameter and its known critical value. Such a comparison can concern only *one* model element of the WRS (arc or node). For non-point sources, the extreme value of the parameter along the arc is compared to the critical values of the same parameter. Some examples of typical elementary events are:
- reservoir drawdown below the minimum drawdown level;
- increased (exceeding MPC) concentration of any pollutant in this reservoir;
- increased river water discharge leading to a disastrous rise of water level in the beginning or the end of a certain river section;
- increased concentration of any pollutant in a fixed point of water withdrawal;
- increased water discharge in any river section, resulting in a water level rise above the assigned mark at least at one site of this section;
- conditions under which at any site within a section the concentration of a certain pollutant (taking into account its input from non-point sources) exceeds the MPC for this particular pollutant.

The values of the elementary events vary with the time, and they can be either *true* or *false* (or, in numerical terms, *one* or *zero*, respectively).

Composite events are situations, arising in a process of WRS operation, the integral parameters of which are of a certain interest to the decision-maker. The DM can create these events using logical operations with elementary events. The *operands* of composite events (elementary events) can refer to different dynamic parameters of one and the same element of the WRS, to different arcs and nodes on graph G, and can also be analyzed at different moments of time. This allows constructing practically any composite event.

Let us consider several examples showing how the composite events can be described on the basis of the elementary ones.
1. Let the control of a reservoir be performed according to certain dispatch control rules. The lines showing time-dependent critical fillings separate the zones on the plot of these rules. Within each zone the management rules do not change. Consider the following two elementary events:
    (a) at a certain moment of time the reservoir filling level mark does not exceed the upper boundary of the zone under consideration;
    (b) at the same moment of time the reservoir drawdown level mark does not fall below the lower boundary of the zone under consideration.
A conjunction of these two events is the composite event exemplifying the application of the reservoir management rules for the zone under consideration.
2. The operation of a hydraulic structure is fault-free during the passage of the maximum flow only if two conditions are satisfied simultaneously: the preserved elevation of the abnormal level mark during the reservoir filling, and the non-exceedance of the maximum discharge capacity of the water-release facility.
3. Normal operation conditions of the reservoir are the ones under which it is filled not higher than its full volume mark, and drawn down not lower than its minimum drawdown level mark, and the concentrations of all the pollutants in it are not exceeding MPC. It is obvious that such conditions can be easily presented as a simple logical expression.
4. In decision-making for a real WRS an additional attribute is assigned sometimes to some water users. This attribute can denote the branch of industry which a user belongs to. Assume that this attribute takes the value denoting hydropower production and compare the flow in the inlet of a respective arc to the water demand of each water user with the above mentioned attribute value. If for all these water users the flow in the inlet of the arc is equal to their water demand then the water demand for hydropower production in the considered river basin or region is fully satisfied.
This example can be extrapolated to arbitrary subsets of WRS elements. Depending on the value of the assigned attribute, we can distinguish the attributes that pertain, for example, only to a certain river, a certain administrative region, territorial unit, etc.

With the help of statistical processing of some dynamic parameters of WRS it is possible to obtain different reliability characteristics traditionally expressed as *calculated probability* in water economy, generalizing the notion of calculated probability. Such generalizations allow considering empirical probabilities of fitting of any dynamic operation parameter into the required area, whereas the respective empirical probabilities can be calculated not only for a one-year period. The necessity in such generalization can arise, for instance, if we try to account for the cumulative negative effect of the repeated interruptions in water supply. Sometimes it is expedient to analyze the reliability of the WRS operation not for the whole simulation period but only for the selected design-based time intervals. The length of these intervals can vary from year to year depending on the values of other dynamic parameters obtained by simulations. Such integral parameters should ensure the convolution time for the momentary values of the composite events. As a result, the decision-maker will obtain *the probabilistic characteristic* of the event. A detailed description of relationships between model parameters, as well as numerous examples of composite events and integral parameters is found in [5].

6. A Global Model of a Decision-Making Support System

The solution of the tasks in a problem block defines the functioning of a water resources complex in a large river basin. Meanwhile, even an optimal solution of every problem block considered, usually is not optimal for the complex water problem as a whole. We may speak about *an acceptable solution* based on a combination of individual solutions for territorial and river units. Let us call these acceptable solutions among the individual optimal variants *the best solution variants*. A *global model (meta-model)* of DMSS is a procedure that selects a solutions sequence for different tasks, subsystems, problem blocks, and the best solution for WRS as a whole. A variant of formalization of such a meta-model is presented below in a generalized form for two problem blocks described earlier, namely - a choice of pollution protection measures p, and dispatch control of reservoirs r. Each block operates on its own systems of territorial and river section units in a large basin; however, the indexes of these units and time periods will be omitted below for simplicity. Each term of the model listed below is a vector as it denotes the whole combination of water resources and economic values linked meaningfully:

W_+ - water volume coming to the water body from outside;

W_- - water volume leaving the territory of an object;

w - water resources available on the territory of an object;

G_+ - pollutant loads coming to an object from outside;

G_- - pollutant loads leaving an object;

G - pollutant loads discharged into rivers on the territory of an object before protection measures;

g - pollutant loads discharged into rivers after protection measures;

P - parameters of different treatment alternatives;

V - volumes and other reservoir parameters on the territory of an object;

U - parameters of the dispatch rules for reservoir control;

Q_r - water discharges from reservoirs and river flows;

S - parameters of cost for WRS as the sum of costs for protection measures S_p and reservoir control S_r;

D is the conditional profit of WRS as the sum of profits from reservoir control D_r and water quality management D_p.

Let the operators B_p and B_r denote the problem blocks of water quality management and reservoir control. These operators link the conditional profit and economic costs for the objects with all the parameters and characteristics given above. Below, fixed source data are grouped in brackets as $(\cdots)^{con}$, information link parameters of the type "input-output" are grouped as $(\cdots)^{put}$, variable parameters - as $(\cdots)^{var}$ and modelling results - as $(\cdots)^{res}$. Then two problem blocks will be presented as a set of simultaneous equations:

$$\{D_r, S_r\} = B_r \left[\left(w, W^+, G, G^+, V\right)^{con}, \ (g)^{put}, \ (U)^{var}, \ \left(Q_r, W^-, G^-\right)^{res} \right] \qquad (33)$$

$$\{D_p, S_p\} = B_p \left[\left(w, W^+, G, G^+, V\right)^{con}, \ (Q_r)^{put}, \ (P)^{var}, \ \left(g, G^-\right)^{res} \right]. \qquad (34)$$

The global optimal variant for water economy should be created based on the solutions of the tasks $(D_r + D_p) \rightarrow \boldsymbol{max}$ and $(S_r + S_p) \rightarrow \boldsymbol{min}$. Such multi-criteria choice is difficult, as it is in principle impossible to adequately compare the estimations for water resources and economy management in various problem fields. How can S_p and S_r be compared if the cost for reservoir control is negligible in comparison to the cost for water quality management but their importance for WRS is quite comparable? How can the experts estimate the relative importance of high water quality or water consumption? Similar problems exist also in other problem blocks of DMSS. Taking into account these considerations, the choice of the global variant for water economy should not be based on the optimization but on the best solution variant identified applying the principle of preferences. However, the choice of variants in the framework of a problem block relies upon optimization with maybe multi-criteria.

The global solution of a complex water economy problem in a large river basin is possible only using the approach of iterative sequences of individual solutions in various problem blocks. The global meta-model searches for the solutions for the problem blocks that yield the best improvement of the present water economy as a whole. It should be noted that iterations are allowed between subsystems and tasks in a problem block but not between the blocks.

7. Remarks on Computer Calculations using DMSS

Computer calculations for subsystems and tasks of DMSS should follow the methodological principles described above. An information system used in DMSS should consist of two parts. First, it is necessary to create a special data bank for each task. The contents and structure of this data bank should suit the requirements of a particular problem task. Second, a special data bank accumulating the interim solutions for different tasks should be created. It will allow performing iterative procedures in a search for the global solution of a complex problem for the whole basin. Some auxiliary programs need to be incorporated in the DMSS structure to allow information exchange between subsystems and tasks of the types of "input-output" and "aggregation-desaggregation". Further, the program codes should contain co-ordination procedures, which generate a sequence of calls for various programs needed on the way to the global solution.

The described methodology is a result of a theoretical generalization of the practical experience with the introduction of computerized calculation routines for different water bodies. The methodology has been applied in practice and showed a possibility of complex global solutions based on multi-variant accounts for concrete objects [1,4]. The created programs are a step on the way of creating a complete DMSS for large water bodies.

8. Conclusions

1. An approach for creating a decision-making support system for optimal management of water resources in large river basins has been presented. This approach is suitable for solution of different tasks and problems and contains a variety of computer codes, data banks and special routines for searching the solutions. The DMSS is formed as a dynamic computer system, operating according to the inquiries. The requirement of resilience is obligatory for all the aspects of the system's development.

2. An effective principle determining the structure of mathematical models in the DMSS have been realized in practice in "two directions" *viz.* individual problems and territory division. As a result, individual subsystems of the DMSS combine the problems of admissible complexity - both from the point of view of their computer realization and input data support, as well as decision-making procedures.

3. The approach for decision-making within the DMSS, presented in the form of the global model (or meta-model), seems to be the only possible way of helping the decision-makers to grasp the multi-problematic character and spatially non-homogeneous structure when dealing with management of large water objects. The necessity in creating specialized data banks follows directly from the structure of the mathematical model suggested for the DMSS. These specialized data banks should include input data banks and banks of results from the solutions of different problems. The composition and problem formulations for water conservation measures consider a broad variety of such measures, their comprehensive effect on water quality, interrelationships between the territories of different sizes and their relation to the description of an admissible ecological state of water resources.

4. A systems analysis approach allows an engineering statement and a mathematical solution of the problem of the dispatch reservoir control. For this purpose a decomposition of the whole complex problem results in the tasks of control of reservoir water resources and the control of flood passage, as well as in a decomposition into optimization and simulation tasks.

5. Some of the described tasks have been applied and tested in practice. When developing software codes for DMSS it is important to create a special information data bank for storage of the intermediate results used in iterative procedures of global solutions.

9. References

1. Demidov, A.V., Korol, V.N. and Levit-Gourevich, L.K. (1997). Principles of Reservoir Management Using Computers. - *Melioratsiya i Vodnoe Khozyaistvo* 3, 37-41 (in Russian).
2. Levit-Gourevich, L.K. (1998). The Problem of Water Reservoir Resources Control. *3rd International Conference on Hydroscience and Engineering ICHE-98,* Cottbus, Germany, CD-ROM.
3. Levit-Gourevich, L.K, Priazhinskaya, V.G. and Yaroshevskii, D.M. (1998). Decision-making Support System for a Large River Basin. *3rd International Conference on Hydroscience and Engineering ICHE-98,* Cottbus, Germany, CD-ROM.
4. Yaroshevskii D.M. (1997). Selection of the Form of Calculated Flood Hydrograph in Designing Water Release Facilities. Destructive Water-Congress'96, June 24-28, 1996, Anaheim, USA (Paper no. D-21) IAHS Press, 41-50.
5. Yaroshevskii, D.M. (1998). Simulation Model for the Estimation of Comprehensive Risk in the Operation of River Basin Water Resource Systems. *3rd International Conference on Hydroscience and Engineering ICHE-98.* Cottbus, Germany, CD-ROM.

CONTROL OF REGIMES OF WATER RESOURCE SYSTEMS WITH STOCHASTIC INDETERMINACY OF THE HYDROMETEOROLOGICAL DATA

A. M. REZNIKOVSKY and M. I. RUBINSHTEIN
Energosetproject Institute
105058, Moscow, ul.Tkatskaya, 1. Russia

1. Dispatch control of the regimes of developed water resource systems

1.1 MAIN CONCEPTS

Developed water resource systems (DWRS) belong to a class of complex dynamic probabilistic systems. The DWRS is called probabilistic because the main resource, water, enters the system in accordance with a probabilistic process. Today, when the accuracy in practical applications is sufficient, river runoff can be considered to be a stochastic process following the known probability distributions. Numerous studies were devoted to the probabilistic relationships of river runoff (e.g. [2, 5, 6])

The requirements imposed on the water resources by several water users and water consumers may be considered with a sufficient accuracy to belong to the "deterministic" category. However, water use for irrigation in zones with insufficient moisture is a probabilistic process depending on several stochastically varying geophysical factors.

To assure a reliable functioning of a DWRS, we need (in terms of operation analysis) a certain set of *active control facilities*. For the DWRS this set is quite small. *reservoirs* are the main example of these. It is possible to use a variety of tools in the DWRS. However, when planning DWRS and optimizing their operation schemes, they are usually not taken into account. Since hydrological forecasts are not sufficiently accurate and not available in due time, special, independent of forecasts, control methods of the regimes of the reservoirs in the DWRS are used when planning and operating the DWRS.

M.V. Bolgov et al. (eds.), Hydrological Models for Environmental Management, 215–225.

1.2. FUNDAMENTALS OF THE CONTROL METHOD

The method of dispatch control for the regime of the reservoirs of the DWRS, in principle, consists of constructing the following type of relationships on the basis of the probability distributions of inflow of water to the reservoirs of the DWRS and the requirements of the water consumers:

$$\alpha_{i,j} = f\left(Z_{i,j}; \Pi_{i,j}\right) \tag{1}$$

where α is the output of the hydroelectric complex; i is the time interval in the calculation; j is the number of elements (reservoirs) in the DWRS; Z_{ij} are indexes characterizing the reserve of water in reservoir j at time i; and $\Pi_{i,j}$ are indexes characterizing the amount of water consumed in time interval i.

These relationships allow assigning, for a set of control parameters and uncertain hydrological input for every moment of time, the output of a hydroelectric complex or a number of such complexes as well as for all the hydroelectric complexes with hydropower stations belonging to one or several linked power systems.

If there is no regulating capacity or if it is insufficient (with daily or weekly regulation), the inflow of water to the reservoir of the DWRS or its short-term forecast for the next days can be used as the control parameters. If there is enough capacity for seasonal and long-term flow control, the water volume in the reservoirs, i.e. the available water resources can be used as the parameter determining the regime of the reservoirs of the DWRS.

A control of the regime of a reservoir in accordance with eq. (1) is called dispatch control [3,6,7,8]. The dispatch control rules should guarantee the closest possible agreement of the performance indexes of the hydropower station for an unknown inflow with the "reference" case. This means they should agree with the indexes characterizing the conditions when the long-term inflow of water to the hydropower stations was known at the time for assigning the output of the hydroelectric complex or a cascade of complexes.

The dispatch rules consist of the rules themselves and the dispatch curves. A dispatch curve shows the dependence of the output of the hydroelectric complex on the water level in the reservoirs. The dispatch rules include the methods for using the dispatch curves, rules for the distribution of the water among the consumers, the distribution of the regulation functions among the hydroelectric complexes, and the distribution of functions among the cascade of hydroelectric complexes in the power system or the power pool.

In complex systems the control of the individual elements should be uniform and, at the same time, it should be independent for several of them. This can be done provided all the active control facilities are assembled in a coordinated *autonomous-hierarchical structure* that can optimize the functioning of the DWRS. When developing the hierarchy of the control structure, the reservoirs are subdivided depending on their physical characteristics into those compensated (i.e. operating completely independently according to the optimal control rules like those in eq. (1)) and those compensatory. In the latter case it may be possible to control the total output of all of the hydroelectric complexes in the system. The total output of all the hydroelectric complexes in the system (α_i) is equal to

$$\alpha_i = \alpha_{i,k} + \sum_{j=1}^{n} \alpha_{i,j} \tag{2}$$

where k is a subscript standing for a compensatory reservoir; n is the total number of compensated hydroelectric complexes in the DWRS and $\sum_{j=1}^{n} \alpha_{i,j}$ is the total output of the hydroelectric complexes being compensated.

The output of a hydroelectric compensatory complex is now obtained as

$$\alpha_{i,k} = \alpha_i - \sum_{j=1}^{n} \alpha_{i,j} \tag{3}$$

where $\alpha_{i,k} = F\left(Z_{i,k}\right)$.

The following three main objectives for the system functioning are defined in the control rules for the reservoirs of the DWRS:
• under catastrophically wet conditions, safety of the structures in the DWRS and the pools of the hydroelectric complexes;
• under dry conditions, guaranteed fulfillment of the demands of the water users and the water consumers with a prescribed level of reliability;
• under normal conditions, the maximal economic effect in all branches of the water economy.

In several cases, in the rules it is possible to combine the control not requiring forecasts with the use of hydrological forecasts with a sufficient accuracy. It is advisable to do this trying to meet the first and third of the above-mentioned objectives.
The main objectives for controlling the regimes of a hydroelectric complex and their cascades in the DWRS determine the composition of the regime zones on the dispatch curve.

218

2. Regime zones in the dispatch curves

The main goal of the dispatch control of the regimes of the hydropower stations is to assure the free capacity and a certain water volume in a reservoir providing its functioning for a certain purpose(s). This is done by subdividing the capacity of the reservoir into corresponding zones (see Figs. 1 a and b).

(a)   (b)

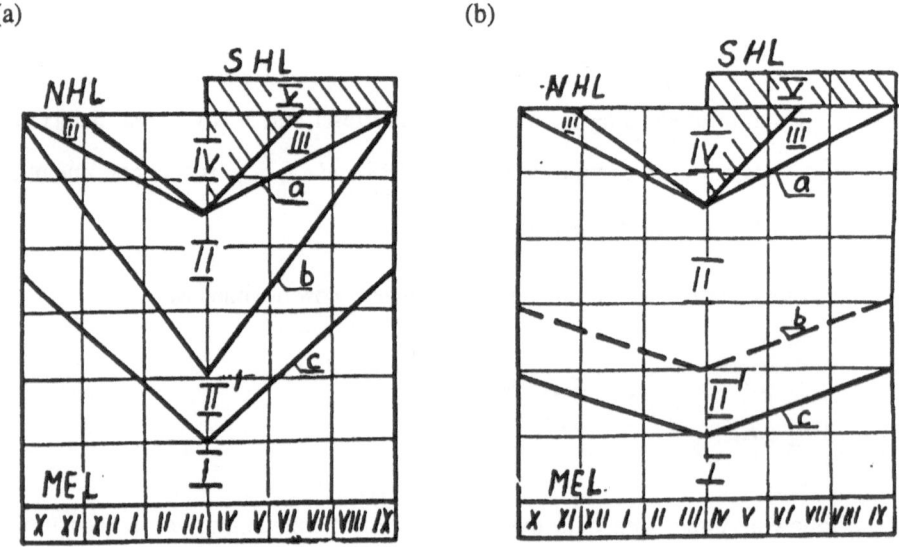

Figure 1. Dispatch curves for the regime of a hydropower station with a seasonal (a) and long-term (b) runoff storage.
Symbols: FHL-forced headwater level; NHL-normal headwater level; a and b - upper and lower bounds of the firm output zone; c - the lower bound of the zone with reduced firm output; I - zone of the minimal permissible output; II-zone of the normal firm energy output; II'-zone of reduced firm energy output; III-zone of energy output exceeding its firm value; IV-zone of maximal water discharge through the hydropower station; V-the maximal water discharge through the hydropower complex. Roman figures in the bottom of the diagrams denote months.

2.1. THE SAFETY ZONE OF THE HYDROELECTRIC COMPLEX

Figures 1 a and b present a schematic diagram of the dispatch curves for operation of the hydroelectric complexes with reservoirs for seasonal (Fig.1 a) and long-term (Fig.1 b) flow control. The safety zone is bounded on the right side by the lower boundary of the zone of the maximal output of the hydropower station (IV) and the normal headwater level (NHL), and on top by the forced headwater level (FHL) (the shaded part of the curves). The method for constructing the safety zone in the dispatch curves, including the zone of maximal output of the hydropower station, is described in detail in [1-4,6-8].

The operation in zone V of the dispatch curve is conditioned on ensuring that in this zone, according to the adopted norms, the normal headwater level (NHL) should not be exceeded in the main design case and the forced headwater level (FHL) in the test case. To satisfy this condition, the water level in a reservoir at the beginning of the flood period should be lowered to a so-called level of forced depletion [8]. The reservoir is filled up to the NHL with maximal flow through the hydropower station, and then to the FHL with maximal flow through the hydroelectric complex, i.e. when the discharge capacity of all the structures (such as the turbines, the weir spans, the bottom discharge, the locks, etc) is used. In cases when the main purpose of the hydroelectric complex is to damp flood peaks, all of the water passages in the hydroelectric complex are opened before the NHL is reached.

2.2. THE ZONE OF ELEVATED OUTPUT OF THE HYDROELECTRIC COMPLEX

The zone of elevated output of the hydroelectric complex or of the energy output of the hydropower station (III) is found on the dispatch curve between the zone of the firm output (II) and the maximal output of the hydropower station (IV) (Fig. 1 a). In the zone of elevated output, this latter quantity can rise directly proportional to the increase in the water level in the reservoir. Alongside with this principle, there are many other methods of controlling the "surplus" output or energy output, that is, an output exceeding the firm one. Applying these methods does not usually lead to a substantial increase in the energy output compared with the alternative of control using the dispatch rules. In the latter case, the energy output of the hydroelectric station differs from the "reference" case by about 1%.

2.3. THE ZONE OF FIRM OUTPUT OF THE HYDROELECTRIC COMPLEX

The water accumulated in this zone (II) is used to provide the firm water discharge or electricity delivery for the system with the adopted design frequency. The firm output of the hydroelectric complex consisting of water resources and power systems can have one or many stages. However, it usually has two stages *viz.* a normal firm stage and a reduced output stage. When the inflow is not known, the firm output can be obtained with a given reliability only if the hydraulic resources found in zones II and II' in Figs. 1 a and b are not utilized during the design dry period with an elevated output. Therefore, the boundaries of the zone of firm output of the hydroelectric complex should be the water levels in the reservoir in the design dry period. The firm (normal and reduced) output is determined under these conditions (see lines a,b,c in Figs. 1 a and b). The calculation of the firm output of the hydroelectric complex and the construction of the corresponding zone on the dispatch curve is discussed in more detail in [1,3,5,6,8]. If there is a cascade of hydroelectric complexes or if the latter are linked, the zone of their total firm output is constructed on the dispatch curve of the compensatory hydroelectric complex. The total firm output of the hydroelectric complexes is found for their joint operation in accordance with the distribution of the regulation tasks among them.

3. **The scheme for controlling the regimes of joint operation of the hydroelectric complexes in the system**

3.1 FUNDAMENTALS OF THE CONTROL SCHEME

At present only two hydroelectric complexes in Russia operate as independent power systems *viz.* the Zeisk and the Kolyma complexes. All the other hydroelectric complexes have cascades of stations or operate together with other hydropower stations in another regional power system or power pool.

Note that the Kolyma hydropower station in the Magadanenergo power system will no longer be a separate station after the construction of the Ust-Srednekansk hydropower station is completed. Also the Zeisk hydropower station in the Vostok pool will operate together with the latter in the power pool after the construction of the cascade of the Bureisk hydropower stations is completed. Thus, in the nearest future there will be no large hydropower stations and hydroelectric complexes in Russia operating separately in power systems.

To ensure a joint operation of hydropower stations and complexes in water resource and power systems, it is necessary, as noted earlier, to develop autonomous-hierarchical dispatch rules for controlling their regimes in order to obtain the maximal possible effect from their operation and safe passage of the flood extremes. For this purpose the regulating functions of the hydroelectric complexes are first established dependent on the hydroelectric characteristics of the stream, topographical characteristics of the reservoirs, their location in the stream (with respect to water consumers and outlets of large rivers), and the demands of water consumers and water users. These functions, in principle, may change during the period under consideration due to the changes in water resources and power systems, i.e. with an appearance of new hydroelectric complexes or consumers.

In the hierarchical structure outlined, the scheme for controlling the regimes of a joint operation of hydroelectric complexes in the system during a certain period can consist of the following:
 • compensated hydroelectric complexes that are controlled independently in accordance with their own dispatch rules;
 • compensatory hydroelectric complexes regulating hydraulic and electrical total energy output of the cascade of stations, which are controlled as a function of the total output of the hydroelectric complexes compensated by them, and the water levels in its own reservoir.

Whenever the water levels in the compensatory reservoir exceed the lower boundary of the zone of maximal output of the hydropower station or of the zone of flood water accumulation (Fig. 1 a and b), the hydroelectric complex is transferred to an autonomous operating regime ensuring its own safety and the safety of its pools during the passage of flood waters.

The zones of elevated output and maximal discharge through the hydroelectric power station on the dispatch curve of the hydroelectric compensatory complex do not differ from similar zones on the dispatch curve of a compensated hydroelectric complex. The zone of firm output is the zone of a total firm output of the hydropower stations in a cascade or a system. The construction of this zone methodologically is similar to the construction of the zone of firm output of an independently operating hydroelectric complex. In this case, however, the output of the hydroelectric compensatory complex is found using eq. (2).

There are cases in which the hydroelectric complex is a compensatory reservoir not only for the energy output of the hydroelectric power stations in its own cascade, but also for the energy output of other cascades in the system that have their own compensatory reservoirs. Such a compensatory reservoir is a compensatory reservoir of the highest degree. When constructing the zone of firm output on its dispatch curve and determining the value of its own energy output according to eq. (2), an account is taken of the energy output of the compensated hydroelectric complexes of its own cascade and of all the hydroelectric complexes of the compensated cascade. The project for a joint operation of the Volga-Kama and the Angara-Yenisei cascades in the United Power System of Russia may serve as an example of such procedures.

3.2. JOINT OPERATION OF THE VOLGA-KAMA AND THE ANGARA-YENISEI CASCADES OF HYDROPOWER STATIONS

The Volga-Kama and the Angara-Yenisei cascades of stations are the largest in Russia. The water power characteristics of the hydroelectric complexes within these cascades and their regulating functions are shown in Tables 1 and 2.

In the Volga-Kama cascade the Volga hydropower station compensates the energy output. The Bratsk hydropower station compensates the energy output of the Angara and the Yenisei hydropower stations, and of the hydropower stations of the entire Volga-Kama cascade (Tables 1 and 2).

TABLE 1. Distribution of the regulating functions among the hydroelectric complexes of the Volga-Kama hydropower cascade

No	Hydroelectric complex	Installed capacity, MW	Dimensionless live storage	Flow control	Regulating function in the cascade of stations	Method of compensation	Area of compensating effect	River
1	Ivankov	30	0.10	annual	compensating	hydraulic	Downstream pool of the Uglich hydropower station	Volga
2	Uglich	110	0.04	seasonal	compensated	—	—	Volga
3	Rybinsk	370	0.40	annual	compensating	electrical and hydraulic	Downstream pool of the Nizhegorod hydropower station	Volga
4	Nizhegorod	520	0.05	seasonal	compensated	—	—	Volga
5	Cheboksary	1400	—	—	compensated	—	—	Volga
6	Kama	504	0.18	annual	compensating	hydraulic	Downstream pool of the Votkinsk and Nizhnaya Kama hydropower stations	Kama
7	Votkinsk	1000	0.07	seasonal	compensated	—	—	Kama
8	Nizhne-Kama	1248	—	—	compensated	—	—	Kama
9	Volga	2300	0.14	annual	compensating	hydraulic and electrical	Downstream pool of the Volgograd hydropower station and energy output of the entire cascade of stations	Volga
10	Saratov	1360	—	daily	compensated	—	—	Volga
11	Volgograd	2540	0.03	weekly	compensated	—	—	Volga
	TOTAL	11390						

TABLE 2. Distribution of the regulating functions among the hydroelectric complexes of the Angara-Yenisei hydropower cascade

No	Hydroelectric complex	Installed capacity, MW	Dimensionless live storage	Flow control	Regulating function in the cascade of stations	Method of compensation	Area of compensating effect	River
1	Irkutsk	660	0.76	long-term	compensating	hydraulic and electrical	Yenisei and Angara hydropower stations	Angara
2	Bratsk	4500	0.50	long-term	compensating	hydraulic and electrical	Yenisei and Angara hydropower stations, Volga-Kama cascade	Angara
3	Ust-Ilimsk	3840	0.03	seasonal	compensated	—	—	Angara
4	Boguchansk	3000	—	seasonal	compensated	—	—	Angara
5	Sayano-Shushensk	6400	0.30	seasonal	compensated	—	—	Yenisei
6	Mainsk	321	—	—	compensated	—	—	Yenisei
7	Krasnoyarsk	6000	0.35	annual	compensating	electrical	Sayano-Shushensk and Mainsk hydropower stations	Yenisei
	TOTAL	24721						

The dispatch rules for a joint operation of hydropower stations enable to benefit from non-synchronicity of the river flow and different regulating characteristics of the reservoirs when the unregulated energy generation by the compensated hydropower stations becomes a component in the total firm output. The waterpower effect from linking the first three hydropower stations of the Angara-Yenisei cascade resulted in a rise of their total firm capacity by 500 MW or 11%. Linking seven of the hydropower stations of the Angara-Yenisei cascade to those of the Volga-Kama cascade can lead to a rise in their total firm capacity by 2180 MW or 16%, and of their firm electricity generation by 12.7 billion kWh or 10%. The storage coefficient for electricity generation for this linkage is 91%. Such a high regulated energy output indicates that the dispatch rules for flow control and, thereby, the energy output of the hydropower stations, provide an optimal control of the functioning of the cascades of hydropower stations and their conglomerates in the power system when the hydrological input is uncertain, i.e. long-term hydrological forecasts are not available in due time and are insufficiently accurate.

In Russia the dispatch rules for control represent an integrated part in "The Main Principles for Rules for Utilizing Water Resources in Reservoirs", which are obligatory for operating hydroelectric complexes. Note in addition that dispatch control of the regime of a hydropower station is necessary whenever it operates together with non-traditional renewable sources of energy (NRSE) and also when solving problems of fuel supply for the power systems.

4. Dispatch rules for a hydroelectric power station operating together with wind power stations

The dispatch rules for flow control allows controlling the regime of a hydropower station when it operates together with power stations utilizing non-traditional renewable sources of energy and, in particular, with wind power stations. The energy generated by these stations is not constant and its level cannot be predicted.

When operating together with hydropower stations having reservoirs with seasonal and long-term flow control, the unregulated energy generated by the wind power stations becomes part of the controlled firm output of the hydropower stations operating together with these latter. In every interval of time, all of the energy generated by the wind- and the hydropower stations, which together form the total firm energy output, should be transmitted to the power system. When the wind power stations are shut down, the hydropower station, the cascade of such stations or all the hydropower stations in the power system have to generate an amount of energy corresponding to the total firm energy output (sometimes partly at the expense of the water that has been saved during the joint operation with the wind power stations).

Determining the level of the total firm capacity of the wind and hydropower stations, the recurrence interval will be the dry recurrence interval when the hydropower station operates with the total firm capacity (for a season, a year, a number of years, depending on the regulating characteristics of the reservoirs). The total firm capacity of the hydro and the wind power stations will be equal to the sum of the firm capacity of the hydropower station and the mean capacity of the wind power stations during the recurrence interval. The increase in the firm capacity in the power system due to the joint operation of the hydro and wind power stations is equal to the mean capacity of the wind power stations during the period when the hydropower station operates with a firm output. The total energy output of the hydro and wind power stations will be controlled using the dispatch rules for operating the hydropower stations as a function of the water levels in the reservoir. If the water levels in the reservoir are such that the hydropower stations can operate with an energy output greater than the total firm energy output, then the hydro and wind power stations operate independently of each other. In this case, the energy output of the wind power stations will be used as seasonal energy. Thus, the regime with a joint operation of hydropower and wind power stations is similar to the regime with a joint operation of compensatory hydropower stations and compensated stations.

5. Dispatch rules for flow control solving the problem of fuel supply

The amount of fuel consumed in the power systems depends on the amount of compensatory energy generated at thermal power stations and used by hydropower stations for energy production. The difference between annual energy generation during wet and dry years at hydropower stations can reach 30% even in power systems with reservoirs for long-term flow control. As the energy generated at a hydropower station for the forthcoming annual or quarterly recurrence interval cannot be determined precisely, the level of fuel supply has to be given as a probability. With this in mind, the energy generated by the hydropower station is calculated for the forthcoming year or quarter of a year on the basis of the entire hydrological series. The calculation of the energy generated at all hydropower stations of a cascade or a system for every water year starts from the actual level of water in the reservoir. Doing this, all of the dispatch rules for controlling the regime of a hydropower station, and interaction between hydropower stations in a cascade and between cascades of stations in the system, which have been adopted for normal operating conditions, are followed. Distribution curves of the annual or quarterly energy generation of the hydropower stations in the system are constructed on the basis of these calculations. These curves serve as the initial data for determining the required fuel supply to the power system in the forthcoming recurrence interval. The method for determining the required fuel supply is presented in detail in [9].

While elaborating the electricity rates, it can be necessary to determine the energy generation at a hydropower station in the power system for some future recurrence interval in accordance with the dispatch rules for operating these stations.

6. Conclusions

1. The autonomous-hierarchical rules of control can be used to ensure optimal functioning of the developed water resources systems in the case of stochastic uncertainty in the hydrometeorological information. This has been proved through a long experience of operating the developed water resources systems in Russia, in the Volga and Yenisei river basins, and also in several countries of the former USSR.

2. When reservoirs for long-term flow control are available, using these rules for the control of the regime of the hydropower station yields an operation regime close to the optimal one for a known river runoff (with an accuracy of up to 1%).

3. The autonomous-hierarchical control rules also allow utilizing the hydrological and waterpower effects from the joint operation of hydroelectric complexes in cascades of stations and in power pools.

4. These control rules allow optimizing the joint operation of hydropower stations and power stations utilizing non-traditional renewable sources of energy (windpower stations, in particular). They allow also determining the design value for the energy production at hydropower stations in order to estimate the required fuel supply to power systems and to elaborate electricity rates.

7. References

1. Asarin, A.E. and Bestuzheva, K.N. (1986) *Vodnoenergeticheskie raschety (Water-Power Calculations)*, Energoatomizdat, Moscow (in Russian).

2. Kritskii, S.N. and Menkel', M.F., (1950) *Gidrologicheskie osnovy rechnoi gidrotekhniki (Hydrological Fundamentals of Hydraulic Engineering)*, AN SSSR, Moscow-Leningrad (in Russian).

3. Kritskii, S.N. and Menkel', M.F., (1952) *Vodokhozyaistvennye raschety (Flow Control Design)*, Gidrometeoizdat, Leningrad (in Russian).

4. Pleshkov, Ya. F., (1972) *Regulirovanie rechnogo stoka (River Runoff Control)*, Gidrometeoizdat, Leningrad (in Russian).

5. Reznikovsky, A.M. (ed.) (1969) *Vodnoenergeticheskie raschety metodom Monte Carlo (Waterpower Calculations by the Monte Carlo Method)*, Energya, Moscow (in Russian).

6. Reznikovsky, A.M. (ed.) (1989), *Gidrologicheskie osnovy gidroenergetiki (Hydrological Foundations of Hydropower Engineering)*, Energoatomizdat, Moscow (in Russian).

7. Reznikovsky, A.M. and Rubinshtein, M.I., (1974) *Upravlenie regimami vodokhranilishch gidroelektrostantsii (Control of the Regimes of Reservoirs at Hydroelectric Power Stations)*, Energiya, Moscow (in Russian).

8. Reznikovsky, A.M. and Rubinshtein, M.I., (1984) *Dispetcherskie pravila upravleniya rezhimami vodokhranilishch (Dispatch Rules for Controlling the Regimes of Reservoirs)*, Energoatomizdat, Moscow (in Russian).

9. Reznikovsky, A.M. and Rubinshtein, M.I., (1997) Energootdacha GES, raschetnaya dlya toplivoobespecheniya energosistem s bol'shim udel'nym vesom GES (The Design Energy Output of Hydroelectric Power Stations for Fuel Supply to Power Systems Having a Large Percentage of Hydroelectric Power Stations), *Gidrotekhnicheskoe Stroitel'stvo* 3, 18-23 (in Russian).

10. Tsvetkov, E.V., Alyabysheva, T.M., and Parfenov, L.G., (1984) *Optimal'nye regimy gidroelektrostantsii v energeticheskikh sistemakh (Optimal Regimes of Hydroelectric Stations in Power Systems)*, Energoizdat, Moscow (in Russian).

RISK MAPPING OF GROUNDWATER CONTAMINATION

W. K. WONG[1], I. KRASOVSKAIA[2], L. GOTTSCHALK[1]
and A. BÁRDOSSY[3]

1. *Dept. of Geophysics, University of Oslo*
   P.O. Box 1022 Blindern N-0315 Oslo Norway
2. *Dept. of Earth Sciences, University of Uppsala*
   Villavagen 16 S-752 36 Uppsala Sweden
3. *Institute of Hydraulic Engineering, University of Stuttgart*
   Pfaffenwaldring 61 70550 Stuttgart Germany

Abstract

In this study a fuzzy rule-based methodology for risk assessment of groundwater contamination in the event of an accidental release of contaminant is described. A degree of risk is considered related to the breakthrough time to the groundwater table. Two spill scenarios are studied: the accidental release of jet-fuel due to leakage from a storage tank and a broken pipe. Compared to traditional modelling techniques, the fuzzy approach does not require a prescribed model structure or detailed information on parameters. Due to the unavailability of appropriate contamination data for deriving the rule system, fuzzy rules are constructed instead based on synthetic training sets. These training sets consist of breakthrough time predictions generated by a flow model utilising the Green & Ampt concept. The simulation results are randomly split into two sets and used for calibration and validation purposes. The resulting rule systems correctly identify approximately 70% of the risk categories. These fuzzy rule models are then applied to the Gardermoen area where detailed site-specific geophysical parameters are not known and hence, fuzzy premises are derived only from expert knowledge. Maps based on risk assessment of the system responses can be made and they are useful tools for planning and risk management purposes.

1.   Introduction

The opening and operation of a new main airport on Norway's largest unconfined aquifer at Gardermoen constitute a potential danger to the local groundwater resources. Stringent injunctions have been imposed on the airport authority in order to protect the groundwater for future use as drinking water. Preventive measures and monitoring programmes have already been implemented, but the possibility of accidental spill of contaminant reaching and contaminating the groundwater can never be completely ruled out.

M.V. Bolgov et al. (eds.), Hydrological Models for Environmental Management, 227–240.
© 2002 *Kluwer Academic Publishers. Printed in the Netherlands.*

Due to the potential attenuating capability, the unsaturated zone may serve as a protecting filter against contamination. Various numerical models, as for example SUTRA (Voss [24]) and FEMWATER (Yeh *et al.* [26]), have been developed and used to study the behaviour of the contaminant in the unsaturated zone. The purpose of modelling contaminant flows is to predict how far and how fast a contaminant will travel in the subsurface. From the protection of groundwater point of view, the breakthrough time to the saturated zone, that is to say the travel time to the groundwater table, is a decisive factor. This information is valuable because it provides the decision-makers and engineers with a time frame within which they decide upon the appropriate remedial measures to be taken. However, the quality of such time predictions will always depend on the knowledge of a specific site. As we have no foreknowledge of where an accident may take place, site-specific hydrogeologic model parameters are therefore not available. Besides, there are several variables which the propagation and migration of contaminant are dependent on, i.e. the amount of contaminant spilled, the time period over which the spill occurred and the area of infiltration of the contaminant – all of which are not known in advance. The substantial uncertainties involved in model description of the interaction of different geophysical, chemical, biological processes further complicate the whole situation. The complexities of these models and demands for high quality field data for the determination of model parameters thus make them less applicable in emergency situations.

Fuzzy rule-based modelling provides a new alternative for dealing with such uncertainties and imprecise information. The main objective of this study is to develop a fuzzy rule-based risk assessment methodology to assist decision-makers in estimating the risks of groundwater contamination associated with different spill scenarios. Comprehensible risk maps based on thorough risk assessments may be powerful tools for remediation planning and risk management purposes.

2.   Definition of risk

Risk may be defined as the potential for unwanted consequences from impending events or a measure of probability and severity of adverse effects (Lowrance [16]), or as the danger that undesirable events represent to human beings, the environment and economic values (Aven [1]). In the context of this study, the accidental release of jet-fuel within the Gardermoen area is identified as the causative event that may lead to risk occurrence, while the contamination of the groundwater resources is defined as the possible consequence of risk exposure. It must be noted that neither causative events nor their possible outcomes alone represent risk unless they are seen in the light of their adverse effects.

The degree of risk is defined by the breakthrough time of jet-fuel in this study, and 7 risk categories have been applied in the risk evaluation (Table 1). The risk of groundwater being contaminated is much higher if the breakthrough time is short. The classification is arbitrary and may therefore be changed when appropriate.

Table 1.   The risk categories applied in this study.

Risk category	Breakthrough time
Very high	Up to 1 hour
High	Between 1 hour and 24 hours
Fairly high	Between 1 day and 7 days
Moderate	Between 7 days and 30 days
Fairly low	Between 1 month and 6 months
Low	Between 6 months and 12 months
Very low	Over 1 year

Since the breakthrough time estimation depends highly on the accident scenario, and so does the risk, two cases will be presented later in this study to illustrate the difference. One of them is leakage from a storage tank and the other represents large spill from a broken pipe.

3.   Fuzzy rule-based modelling

To avoid the difficulties as described previously, the fuzzy rule-based approach is introduced to provide a better alternative. It has already been used in hydrology and other scientific disciplines with satisfactory results (e.g. Bárdossy et al. [2]; Bárdossy [4]; Bárdossy & Duckstein [5]; Pongracz et al. [21]). A fuzzy approach does not require a prescribed model structure or complex equations for describing the interconnections between the input variables and the corresponding responses. Detailed information on various parameters is therefore not necessary. The relationship between the inputs and the responses will on the other hand be established and maintained by a specific rule system. Training datasets are needed in this connection for deriving the appropriate rule system. Expert knowledge can at the same time easily be taken into consideration and incorporated in the rule formulation. The ordinary training set consists of a large number of observations. In the absence of observed contamination data, synthetic training sets, that is to say simulated datasets, are used instead. The model used to generate the training sets in this study is just an example and can be replaced with any other suitable flow and transport model. It is chosen mainly because of its simple model formulation and computational speed.

3.1.   FORMULATION OF THE BREAKTHROUGH TIME MODEL

A quasi-two-dimensional, distributed, single phase flow model inspired by the Green & Ampt concept (Green & Ampt [10]) has been developed for modelling the breakthrough time of jet-fuel to the groundwater. The model focuses only on the flow of jet-fuel. The other processes that may have taken place in the subsurface such as redox reactions, biodegradations, etc. are therefore not taken into account. The jet-fuel is regarded to be immiscible with water. The displacement of fluids in this approach is assumed as piston flow, i.e. one fluid replacing the other, and hence soil water is assumed to be immobile compared with jet-fuel. As a result, a sharp front separating the saturated and unsaturated condition is assumed. The modelling of the initial infiltration process is equivalent to distributing the jet-fuel over a bundle of uniform diameter tubes with homogeneous soil properties and each tube represents an individual process. Jet-fuel is first injected into one of the tubes, and the model calculates the amount of jet-fuel the tube manages to take during the infiltration period. The rest of the jet-fuel is then immediately distributed to a second tube, then a third, and so forth, until the whole volume of spilled jet-fuel is completely infiltrated. This implies no ponding occurs on the soil surface. The downward movement of jet-fuel in each tube is driven by gravity alone. There is no communication between the tubes after the initial infiltration so lateral spreading is ignored.

Several modelling attempts on other organic chemicals using a similar approach have been reported in the research literature. Despite the simple model formulation, it can be concluded that this sharp-interface approach can describe the main features of the infiltration behaviour of chemicals in the unsaturated zone (e.g. El-Kadi [8]; Guigard et al. [11]; Kessler & Rubin [15]; Reible et al. [23]).

The formulation of this model is partly based on Bárdossy & Disse [3]. The general infiltration process of water can be described by the basic Green & Ampt equation which can be written in the following form:

$$f(t) = K_s \left(1 + \frac{H_0 + H_f}{Z_f} \right)$$

(1)

where

$f(t)$ actual infiltration rate at time t;
K_s saturated hydraulic conductivity;
H_0 depth of ponded water on soil surface;
H_f capillary suction at wetting front;
Z_f actual depth of wetting front.

The capillary suction H_f can be obtained from Mein & Larson [17] as

$$H_f = \int_{K_{r,i}}^{K_{r,a}} \psi(K_r) dK_r$$

(2)

where K_r is the relative hydraulic conductivity with $K_r = K(\psi) / K_s$; here index i is used for initial values and index a for actual values.

The relationship between pressure head ψ and water content θ is given by Brooks & Corey [6] as

$$\frac{\theta - \theta_r}{n - \theta_r} = \left(\frac{\psi_a}{\psi} \right)^{\lambda}$$

(3)

where n is porosity, θ_r is residual water content and λ is the grain-size distribution index.

In the mean time, the supply of water stops at the end of the infiltration period. The further advance of the wetting front is under unsaturated condition and (1) becomes:

$$f(t) = K(\theta)\left[1 + \frac{H_f(\theta)(\theta_a - \theta_i)}{Z(t)}\right]$$

(4)

where

θ_a actual water content;

θ_i initial water content;

$Z(t)$ cumulative infiltration at time t and $Z(t) = Z_f(\theta_a - \theta_i)$.

According to Mualem [18], the relative hydraulic conductivity $K(\psi)$ can be represented as

$$K_r(\psi) = \frac{K(\psi)}{K_s} = \left(\frac{\psi_a}{\psi}\right)^\beta$$

(5)

with ψ_a the air entry pressure head and β the Mualem parameter which can be expressed as $\beta = (2+3c)$ where c is a constant.

All these equations are originally meant for water flow in soils, but here they are assumed to be applicable for jet-fuel after the necessary modifications. Saturated hydraulic conductivity is substituted by the effective jet-fuel conductivity while the relative hydraulic conductivity is replaced with relative jet-fuel conductivity. The jet-fuel content can in the same way take the place of the water content in the equations.

The effective jet-fuel conductivity can be calculated via intrinsic permeability with the following equations (e.g. Hillel [12]):

$$k = \frac{K_{s,w}\mu_w}{\rho_w g}$$

(6)

$$K_{s,j} = kK_{r,j}\frac{\rho_j g}{\mu_j}$$

(7)

where k is intrinsic permeability, μ is dynamic viscosity, ρ is density, g is gravity, $K_{r,j}$ is relative jet-fuel conductivity; here index w denotes water and index j jet-fuel.

The retention capacity for the jet-fuel and soil is an important parameter in determining the initial propagation of the spill since migration can only proceed if the retention capacity is exceeded. It is defined as the threshold below which jet-fuel is unable to move (de Pastrovich et al. [19]). This parameter and the law of mass conservation have also been included in the model.

3.2. SIMULATION SETUPS

The simulations are concentrated on two spill scenarios with different initial conditions. A leakage implies a relatively low infiltration rate over a long period of time and the release is thus only confined to a small area. Spill from a broken pipe, on the contrary, means a much higher intensity and involves a larger area. In keeping the total number of simulations to a manageable size, the degree of freedom of the model has been limited and hence, only 8 of the model parameters are allowed to vary. They are all site-specific and mostly related to the physical properties of the soils. The rest of the parameters are considered as constants with their values assigned based on the studies of Pedersen [20] and Wong & Engen [25]. These previous studies also provide the basis for the determination of the possible initial values for the 8 parameters. In addition, the volume of spilled jet-fuel is set to be $11m^3$ for both scenarios in all the simulations. This is equivalent to the storage capacity of a railway carriage. For the simulations describing leakage scenario, the release is assumed to last for three hours while the spill is supposed to be over in just two minutes for broken pipe cases. These values are taken from Holtebekk et al. [13]. Table 2 lists the applied model parameter values and simulations are carried out by the combination of all these values.

From the simulations breakthrough time predictions are obtained, which are in fact the average of 10 similar simulations using different effective jet-fuel conductivity values. Ten simulations are carried out instead of one to compensate for the effect of the heterogeneity of the soils. The conductivity values are quantiles taken from log-normal distributions with their standard deviations equal to the corresponding mean values. These mean values are stated in Table 2.

Table 2.   The parameter values applied in the simulations.

Parameter	Value
Grain-size distribution index, λ	0.3; 0.5; 0.7; 0.9; 1.1; 1.3; 1.5
Mualem parameter, β	1.0; 1.1; 1.2; 1.3; 1.4; 1.5
Air entry pressure head, ψ_a [m]	0.05; 0.2; 0.4; 0.6; 0.8; 1.0
Initial water content, θ_i	0.015; 0.05; 0.1; 0.15; 0.2
Mean effective jet-fuel conductivity, $K_{s,j}$ [m/s]	4.540E-5; 1.234E-4; 3.355E-4; 9.119E-4; 2.479E-3; 6.738E-3
Depth of the groundwater level [m]	3; 4; 5; 6; 7; 9; 11; 13; 15; 20; 25; 35
Porosity, n	0.25; 0.3; 0.35
Retention capacity [l/m³]	5.0; 7.5; 10.0
Residual jet-fuel content	0
Kinematic viscosity for jet-fuel at 5°C, $v_j = \mu_j / \rho_j$ [m²/s]	2.25E-6
Kinematic viscosity for water at 5°C, $v_w = \mu_w / \rho_w$ [m²/s]	1.52E-6
Volume [m³]	11
Release time [s]	120 for leakage; 10800 for broken pipe

The maximum breakthrough time is introduced in the model and set to 3.5E+7 seconds, approximately 400 days. When this value is exceeded, the simulation stops regardless of the actual position of the contaminant front. It is adopted in order to avoid the few cases where the advance of the contaminant front stops before reaching the predefined depth and the simulation runs into an infinite loop.

The simulation results are first grouped according to the two spill scenarios and then randomly split into two sets. One is used as a training set for calibration of the rule system and the other for validation purposes.

3.3. BRIEF REVIEW OF THE DEFINITION OF FUZZY NUMBER AND FUZZY RULE

A fuzzy number A_i consists of pairs in the form $(x, \mu_{Ai}(x))$ where x is an element of some type of continuous set and $\mu_{Ai}(x)$ is the corresponding membership function whose value varies in the closed interval $[0,1]$. This value is dependent on the extent to which the element x belongs to that particular set. Moreover, the membership function must have an increasing and a decreasing part. The simplest fuzzy numbers are triangular numbers represented as $(a_1, a_2, a_3)_T$ with $a_1 \leq a_2 \leq a_3$ (Figure 1).

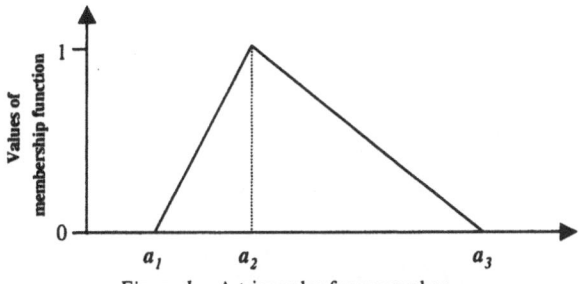

Figure 1. A triangular fuzzy number.

A fuzzy rule consists of a set of premises $A_{i,k}$ in the form of fuzzy numbers with membership functions $\mu_{Ai,k}$ and a consequence B_i also in the form of a fuzzy number. The simplest rules are formulated by using **AND**-operator only:

$$\text{IF } A_{i,1} \text{ AND } A_{i,2} \text{ AND } ... \text{ AND } A_{i,k} \text{ THEN } B_i \tag{8}$$

with i = number of rules and k = number of premises.

In contrast to ordinary, non-fuzzy, crisp rules, fuzzy rules allow partial and simultaneous fulfilment of rules. The degree of fulfilment (DOF) indicates the applicability of the fuzzy rule and can be calculated by using the product inference method:

$$\nu (A_1 \text{ AND } A_2) = \mu_{A_1}(a_1)\mu_{A_2}(a_2) \tag{9}$$

Different fuzzy rules with different consequences can also be applied to the same premises. As a result, the response derived from a rule system is a combination of several individual rule responses. The combination method used in this study is the normed weighted sum combination. The resulting consequence B is based on the responses (B_i, ν_i) where ν_i is the DOF of rule i. The method can be expressed as:

234

$$\mu_B(x) = \frac{\sum_{i=1}^{n} v_i \beta_i \mu_{B_i}(x)}{\max_u \sum_{i=1}^{n} v_i \beta_i \mu_{B_i}(u)}$$

(10)

where n = number of rules and

$$\frac{1}{\beta_i} = \int_{-\infty}^{+\infty} \mu_{B_i}(x) dx$$

(11)

It is often necessary to replace the fuzzy consequence B with a single crisp consequence b. This procedure is called defuzzification. When the normed weighted sum combination method (10) is used for the calculation of the fuzzy consequence, the defuzzified answer can be expressed in the following form:

$$b = M(B) = \frac{\sum_{i=1}^{n} v_i M(B_i)}{\sum_{i=1}^{n} v_i}$$

(12)

If B_i is a triangular fuzzy number with $B_i = (b_{i,1}, b_{i,2}, b_{i,3})_T$, then $M(B_i)$ becomes as simple as:

$$M(B_i) = \frac{b_{i,1} + b_{i,2} + b_{i,3}}{3}$$

(13)

More details on fuzzy set theory or the fuzzy rule-based approach can be found in e.g. Dubois & Prade [7] and Bárdossy & Duckstein [5].

3.4. DETERMINATION OF PREMISES AND RESPONSES

It is important to identify the premises among the input variables that explain the responses best. After a sensitivity test of the model parameters, the first 6 site-specific model parameters in Table 2 are found to be the most important ones for the progression of the contaminant front, namely λ, β, ψ_a, θ_i, $K_{s,i}$ and groundwater level. They are therefore used as premises in the fuzzy rule system. The whole interval of possible premise values of each premise is divided into 12 overlapping classes based on the arranged data from training sets. A membership function is assigned to each of the classes which forms a fuzzy number. Most of them are triangular fuzzy numbers. Figure 2 shows the schematic divisions of the premise values into classes, or fuzzy numbers. The response variable, breakthrough time prediction, is also defined in a similar way. Since the variation of the breakthrough time is of several orders of magnitude, all the estimates are first log-transformed before the classifications.

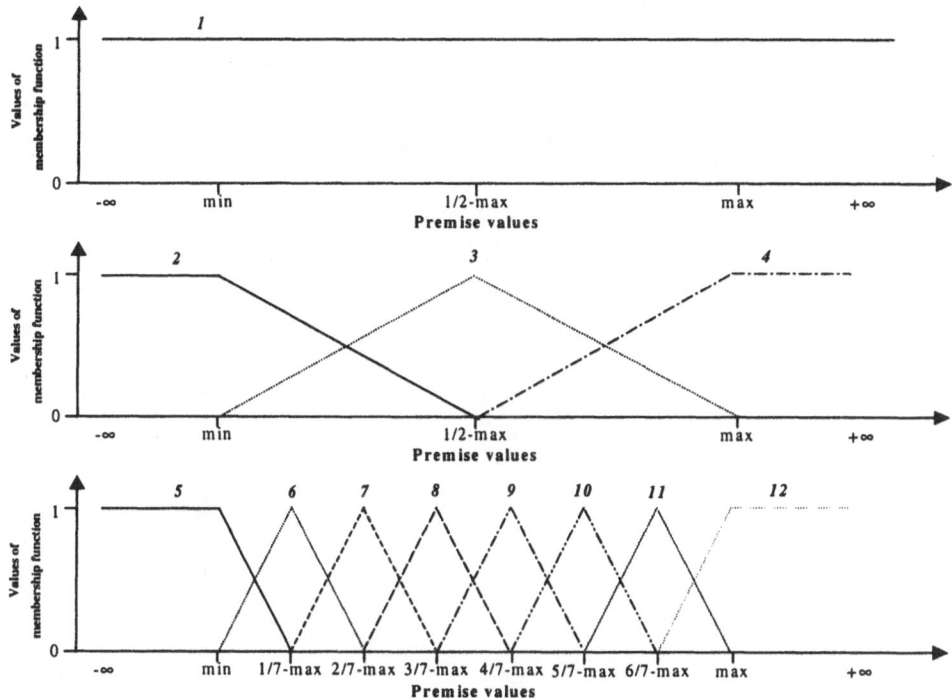

Figure 2. Fuzzy numbers or classes defined in the premises; the italic number indicates the defined class number.

3.5. CONSTRUCTION AND VALIDATION OF RULE SYSTEMS

After the determination of fuzzy numbers in premises and response, the trial and error approach is often used to find an appropriate rule system. Unfortunately, this is a very time-consuming exercise even in the cases with a few premises. In this study, an automatic optimisation routine using the simulated annealing method (e.g. Press *et al.* [22]) is applied to derive the optimal set of rules. This optimisation method minimises the objective function, or error function, which is important for the performance of the rule system. The objective function used in this study is as follows:

$$\sum_{i=1}^{n} \left(\frac{T_{sim}(i) - T_{obs}(i)}{\frac{1}{2}\left(T_{sim}(i) + T_{obs}(i)\right)} \right)^2 \tag{14}$$

where n is the number of observations, T_{sim} is the simulated response and T_{obs} is the observed response.

When the 'observed' breakthrough times are small, this objective function generally offers more accurate simulated responses. Two rule systems are derived, one for each scenario. The resulting rule systems consist of 40 rules each, which are assumed to be independent. They are then validated by other datasets to see if they can sufficiently describe the complex subsurface flow.

236

As the responses provided by the rule systems are log-transformed breakthrough time estimates, they need to be converted back to the original time domain before any further assessment can be made. The responses are in addition classified according to the risk categories which have been defined in Table 1. The analyses show that the overall performances of the two rule systems are quite good. On an average they are able to correctly identify about 70% of the risk categories (Table 3).

Table 3.   The identification of risk categories using fuzzy rule systems.

Risk category	Tank leakage scenario				Broken pipe scenario			
	Correct identification	Over-estimation of risk	Under-estimation of risk	Success rate in %	Correct identification	Over-estimation of risk	Under-estimation of risk	Success rate in %
Very high	4810	-	691	87.4	5	-	20	20.0
High	5706	152	849	85.1	513	3	373	57.7
Fairly high	3271	481	1173	66.4	695	80	432	57.6
Moderate	1795	642	781	55.8	550	126	291	56.9
Fairly low	6711	1954	240	75.4	3551	985	441	71.3
Low	2268	3921	576	33.5	4275	2949	981	52.1
Very low	5188	3995	-	56.5	20271	6113	-	76.8
Total	29749	11145	4310	65.8	29860	10256	2538	70.0

For broken pipe scenarios, the very low success rate of identifying very high risk is due to the lack of training data from such an event; only 15 cases have been used for calibration, which can hardly be regarded as adequate. This may also contribute to the limited success of identifying the succeeding risk categories.

For situations involving small leakage, the picture is slightly different. The rule system performs well with over 85% correct identification of the high risk categories. However, this success rate drops dramatically to only 34% for the category of low risk. It is probably a result of the objective function employed, which tends to underestimate the breakthrough time and hence overestimate the risk.

For both scenarios, a major part of the errors comes from the difficulties in differentiating the risk categories between low and very low. This can partly be explained by the adoption of maximum breakthrough time - an arbitrary limit that has an influence on the determination of response classes and subsequently the rule construction. The objective function also gives rise to another error: it tolerates deviation of large breakthrough time estimates. As the consequences of misidentifying these two categories are probably rather small, they can be put together as one. The success rate of identifying the correct risk categories is significantly improved, to more than 85% on average for broken pipe scenarios and 74% for tank leakage scenarios, which can be considered as very good.

4. Application of fuzzy rule-based model to the Gardermoen area

The Gardermoen aquifer is formed by one of the largest glaciofluvial deltas in Norway. The thickness of this unconfined aquifer varies between 1 and 35 m. The upper part of the aquifer, 20-30 m, is mostly composed of sand underlain by glaciomarine and marine silts and clays (Jørgensen & Østmo [14]). A classification of the sediment deposits in the Gardermoen area based on a 100x100 m grid is shown in Figure 3 (Wong & Engen [25]). As we can see, the whole area is dominated by glaciofluvial deposits. Eolian sand dunes can also be noticed in the northern part of the area while glaciolacustrine deposits can be found in the east. These deposits with rather different soil properties may have different effects on the migration of jet-fuel and hence, the breakthrough time. As detailed information about the soil properties is not available in the studied area, fuzzy premises have to be derived from expert opinion and the Quaternary map. The local groundwater level is also described as a fuzzy premise due to its natural fluctuation. The interpolated groundwater levels of the studied area provided by Engen [9] are fuzzified and applied. Figure 4 summarises the fuzzy premises applied in the modelling.

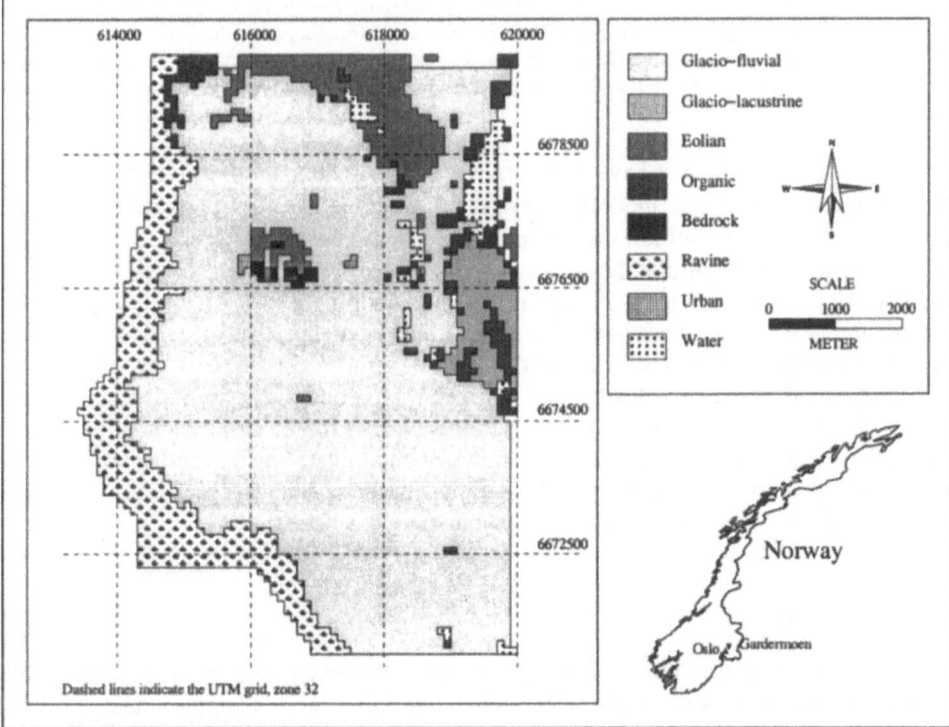

Figure 3. The classification of sediment deposits in the Gardermoen area (modified after Wong & Engen [25]).

238

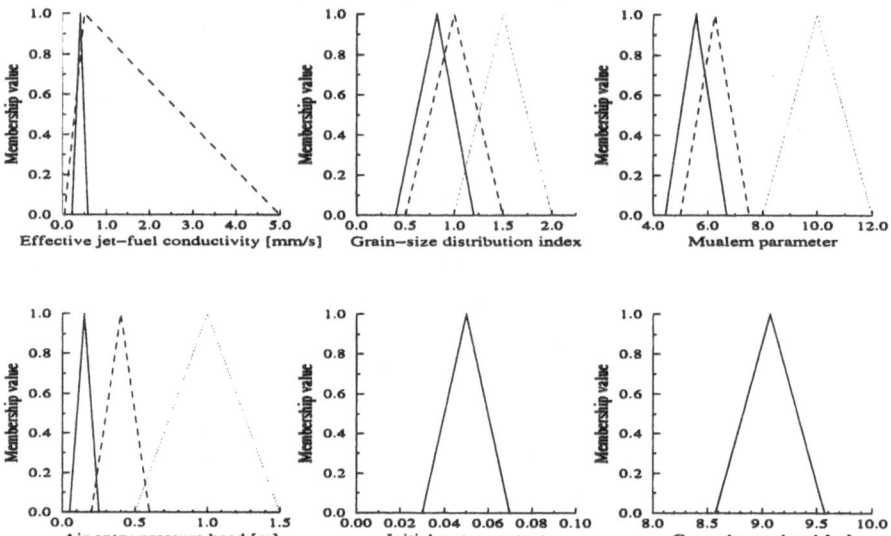

Figure 4. Fuzzy premises applied in the modelling of breakthrough time to the Gardermoen area. The grey solid lines indicate the fuzzy premises for glaciolacustrine deposits, the black dashed lines for eolian deposits while the black solid lines are for glaciofluvial deposits. For the premises initial water content and groundwater level, the same fuzzy number is applied for all the deposit materials.

The same rule systems that have been derived and validated are used for the modelling of breakthrough time. The modelling results are assessed and then presented in form of risk maps. The grid definition and co-ordinate system applied in Figure 3 are also used in these risk maps. Figure 5 shows that when all other conditions are kept the same, a small leakage of jet-fuel from storage tank over a long period of time actually poses a greater threat to the groundwater than a large instantaneous spill caused by a broken pipe. For the broken pipe scenario, over 90% of the area is classified as low or very low risk. It means the jet-fuel will probably not reach the groundwater in 6 months time if a spill really takes place at these locations. However, if the same amount of jet-fuel is allowed to leak to the ground slowly, it will contaminate the groundwater within a month at almost 30% of the locations. The areas in the north and north-east with rather high groundwater level are especially vulnerable.

5. Conclusion

In contrast to the traditional modelling technique, the fuzzy rule-based modelling has shown its ability to deal with uncertainties in this study. Complex mathematical equations are substituted by simple rule formulations while expert knowledge or opinion can also be included in a similar way.

Fuzzy rule systems derived from the synthetic training sets are used to estimate the breakthrough time of jet-fuel. As we are most concerned about the risk of groundwater contamination and not the absolute breakthrough time predictions, the time estimates are classified according to predefined risk categories. The validation results show that the rule systems successfully identify about 70% of the risk categories.

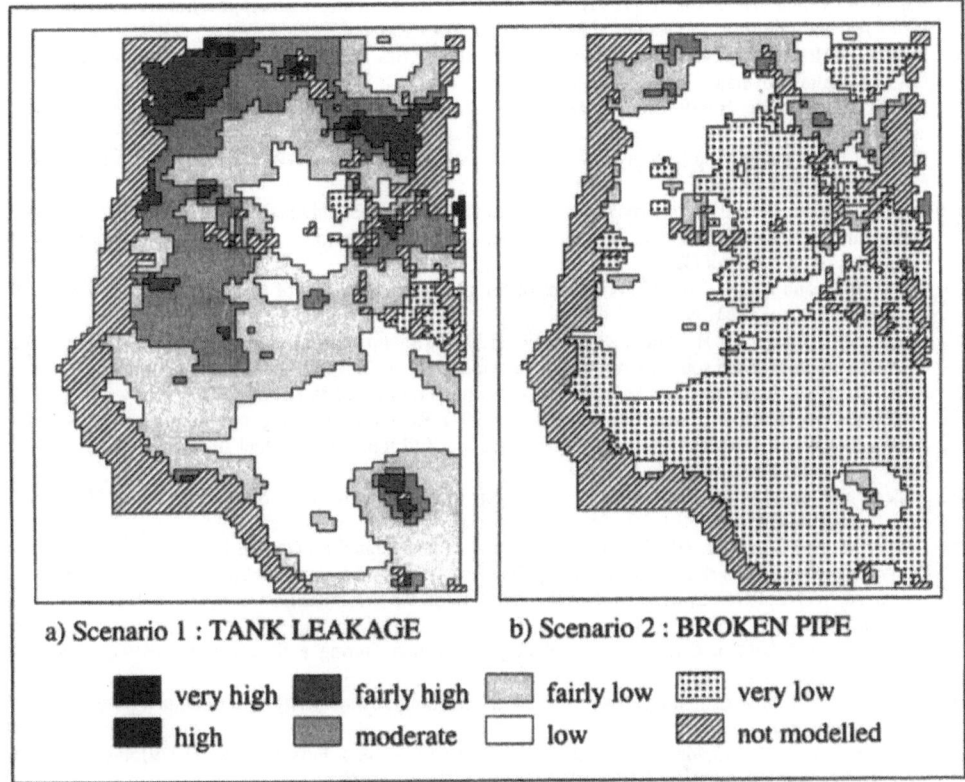

a) Scenario 1 : TANK LEAKAGE   b) Scenario 2 : BROKEN PIPE

■ very high	■ fairly high	▫ fairly low	▦ very low
■ high	▪ moderate	▫ low	▨ not modelled

Figure 5. Fuzzy rule-based risk assessment of two spill scenarios. The risk categories applied are taken from Table 1.

The results from application of the fuzzy rule-based model to the Gardermoen area show that there is a significant difference in risk between the two spill scenarios. In 30% of the locations, when a small leakage occurs there, the jet-fuel will reach the groundwater within 30 days. On the other hand, the quality of groundwater for 90% of the area will not be affected if a large volume of jet-fuel is spread over the area due to a broken pipe, at least not for the first 6 months. This kind of information visualised in the form of a comprehensible map may be very useful for decision-makers when they have to select the most appropriate remediation scheme among many different alternatives.

6.   References

1.   Aven, T. (1992) *Reliability and Risk Analysis*, Elsevier Applied Science, London.
2.   Bárdossy, A., Bogardi, I. and Duckstein, L. (1990) Fuzzy Regression in Hydrology, *Water Resources Research* **26**, 1497-1508.

240

3. Bárdossy, A. and Disse, M. (1993) Fuzzy Rule-Based Models for Infiltration, *Water Resources Research* **29**, 373-382.
4. Bárdossy, A. (1994) The use of fuzzy rules for the description of elements of the hydrological cycle, *Ecological Modelling* **85**, 59-65.
5. Bárdossy, A. and Duckstein, L. (1995) Fuzzy Rule-Based Modeling with Applications to Geophysical, Biological and Engineering Systems, CRC Press Inc., Boca Raton.
6. Brooks, R.H. and Corey, A.T. (1964) Hydraulic Properties of Porous Media, Hydrology Paper 3, Colorado State University, Fort Collins, Colorado.
7. Dubois, P. and Prade, H. (1980) *Fuzzy Sets and Systems : Theory and Applications*, Academic Press Inc., San Diego.
8. El-Kadi, A.I. (1992) Applicability of Sharp-Interface Models for NAPL Transport : 1. Infiltration, *Groundwater* **30**, 849-856.
9. Engen, T. (1995) Stokastisk interpolasjon av grunnvannspeilet i isranddeltaet på Gardermoen (Stochastic interpolation of groundwater table in the ice-contact delta at Gardermoen), Master thesis, Department of Geophysics, University of Oslo, Oslo.
10. Green, W.H. and Ampt, G.A. (1911) Studies of soil physics: I. The flow of air and water through soils, *J. Agricultural Science* **4**, 1-24.
11. Guigard, S., Stiver, W.H. and Zytner, R.G. (1996) The infiltration and movement of immiscible chemicals in unsaturated soil, *Environmental Technology* **17**, 1123-1130.
12. Hillel, D. (1982) *Introduction to Soil Physics*, Academic Press Inc., San Diego.
13. Holtebekk, H., Roald, T. and Sagvolden, T. (1997) Risiko ved lasting og lossing av jet-fuel fra jernbanevogner på Gardermobanen (Risk when loading and unloading jet-fuel from railway carriage on Gardermobanen) Scanpower AS, Norway.
14. Jørgensen, P. and Østmo, S.R. (1990) Hydrogeology in the Romerike area, Southern Norway, *Nor. Geol. Unders. Bull.* **418**, 19-26.
15. Kessler, A. and Rubin, H. (1985) On the simulation of unsaturated oil flow in soils, *Scientific Basis for Water Resources Management (Proceedings of the Jerusalem Symposium). IAHS Publ.* **153**, 195-205.
16. Lowrance, W.W. (1976) *Of acceptable risk : science and the determination of safety*, William Kaufman Inc., Los Altos.
17. Mein, R.G. and Larson, C.L. (1973) Modelling Infiltration During a Steady Rain, *Water Resources Research* **9**, 384-394.
18. Mualem, Y. (1976) A new model for predicting the hydraulic conductivity of unsaturated porous media, *Water Resources Research* **12**, 513-522.
19. de Pastrovich, T.L., Baradat, Y., Barthel, R., Chiarelli, A. and Fussell, D.R. (1979) Protection of groundwater from oil pollution, CONCAWE report no. 3/79, Den Haag.
20. Pedersen, T.S. (1994) Væsketransport i Umettet Sone – Stratigrafisk beskrivelse av toppsedimentene på forskningsfeltet, Moreppen og bestemmelse av tilhørende hydrauliske parametre, del I og II (Flow in the unsaturated zone – stratigraphic description of the top sediments at Moreppen field site and the determination of hydraulic parameters, part I and II), Master thesis, Department of Geology, University of Oslo, Oslo.
21. Pongracz, R., Bogardi, I. and Duckstein, L. (1999) Application of fuzzy rule-based modeling technique to regional drought, *Journal of Hydrology* **224**, 100-114.
22. Press, W.H., Teukolsky, S.A., Vetterling, W.T. and Flannery, B.P. (1992) *Numerical Recipes in Fortran, Second Edition*, Cambridge University Press, 436-448.
23. Reible, D.D., Illangasekare, T.H., Doshi, D.V. and Malhiet, M.E. (1990) Infiltration of Immiscible Contaminants in the Unsaturated Zone, *Groundwater* **28**, 685-692.
24. Voss, C.I. (1990) A finite-element simulation model for saturated-unsaturated, fluid-density-dependent groundwater flow with energy transport or chemically-reactive single-species solute transport (two-dimensional version). U.S. Geological Survey, Reston.
25. Wong, W.K. and Engen, T. (1997) Sårbarhetsstudie av Gardermoen-akviferen I. Modellering av gjennombruddstid av ulike avsetningsmaterialer på Gardermoen området (Vulnerability study of Gardermoen aquifer I. Modelling of breakthrough time of various sediment deposits in the Gardermoen area), The Gardermoen Project, Report series B, No. 14, University of Oslo, Oslo.
26. Yeh, G.T., Lin, H.C.J., Richards, D.R., Cheng, J.R., Cheng, H.P. and Jones, N.L. (1996) FEMWATER: A Three-Dimensional Finite Element Computer Model for Simulating Density Dependent Flow and Transport, Technical Report HL-96.

ECONOMIC RISK OF FLOODING: A CASE STUDY FOR THE GLOMMA RIVER, NORWAY

LARS GOTTSCHALK[1] , IRINA KRASOVSKAIA[2] and
NILS ROAR SÆLTHUN[3]
[1]Dep. of Geophysics, University of Oslo
P.O. Box 1022 Blindern, N-0315 Oslo, Norway
[2]Dept. of Earth Sciences, Hydrology, University of Uppsala
Villavagen 16, S-752 36 Uppsala, Sweden
[3]Norwegian Institute for Water Research (NIVA),
P.O.Box 173 Kjelsås, N-0411 Oslo

1.   Introduction

Recurring extreme floods during recent decades [50] make many believe that the flood danger is greater today than ever before. Indeed, floods (and drought) cause more damage and kill more people than any other natural disaster [37,25]. In what way has the situation become worse? Is it a result of more frequent floods with higher magnitude or is it the vulnerability of our society to floods that has increased?

The hydrologist, by tradition, had a rather narrow perspective on extreme flood events usually assessing the problem in terms of its return period. Only recently has the risk concept come into focus including, as well as the measure of probability, some estimate of the consequences. Approaches based on hazard (water depth and velocity), economic, social and environmental aspects have become more and more common in the design of flood protection [16,19,7,8].

There exist different definitions of risk in hydrology. Risk can be regarded as identical to the probability of exceedance of a certain flood level within a given design period [15,20]. A more common interpretation is the probability of failure of a system (storm sewer networks, flood dikes, bridges) with consideration of all uncertainties involved in the estimation of this [43,38,45,17,34]. The expected damage or the expected loss are often treated as a separate concept. Plate [34] calls this the "generalized risk". Finally, a

M.V. Bolgov et al. (eds.), Hydrological Models for Environmental Management, 241–263.
© 2002 *Kluwer Academic Publishers. Printed in the Netherlands.*

more generally accepted definition of risk is the expected loss, damage or utility [4,2] used also in hydrologic applications [1,32,28,44,42]. Herein, only this latter definition will be used with a special focus on uncertainty.

The emphasis in European efforts to cope with flooding is put on the human factor as the main cause for the increased flood risk manifested in a changing climate and changing environment (land use). This study was carried out in the frame of the Norwegian HYDRA project, which also focused on the human factor with the working hypothesis that "the sum of all human impacts in the form of land use, regulation, flood-protection etc., may have increased the risk of floods". The frequency of an undesirable event and a measure of the severity of adverse effects of it [30] or the danger that undesirable events represent to human beings, the environment and economic values [2] are both parts of the concept of risk. From this point of view, the risk for flooding has obviously increased. More difficult is it to give a unique answer to why it has increased, as such an answer would certainly depend on the individual conditions for each country or even a drainage basin. Here we will try to identify factors that determine the risk of flooding and develop a risk model for the Glomma River, Norway with a basin area of 40000 km^2. This model, for evaluating the economic risk connected to floods, is then applied to the Grue and Åsnes (Fig. 1) municipalities upstream of Nor in the Glomma river basin.

2. Factors influencing the risk of flooding

Factors influencing the risk of flooding may be many. Natural causes (like climate change or variability, land subsidence etc.), growth of population and increasing exploitation of flood plains, land-use change, shortcomings in communication of the flood risk message to decision makers and broad public and also better and more frequent reporting of flood damages are often named as such factors [e.g. 18,16].

There is already a vast literature on possible changes of runoff due to a changing climate or its variability. Krasovskaia [21] and Gottschalk and Krasovskaia [10] give extensive reviews related to Scandinavian conditions. In relation to floods they conclude that while the magnitude of individual floods is not influenced to any significant degree, there is a tendency for a growing number of flood events with increasing temperature in Southern Scandinavia. The overall effect of this is that the probability for extreme

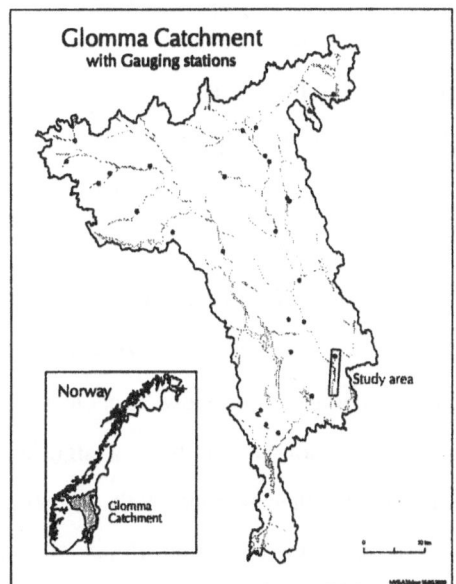
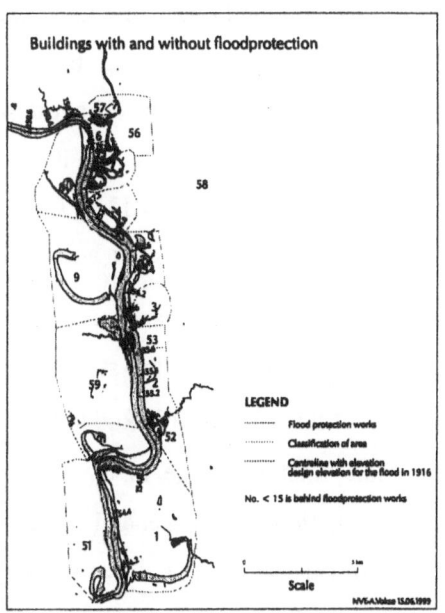

Figure 1. The Glomma drainage basin with hydrological gauging stations (left). The study area around Grue and Åsnes municipalities is marked by a rectangle and shown in detail (right). 18 settlement clusters are identified, of which eight are protected by dikes (bold lines).

events slightly increases with a higher temperature. The increase of the number of flood events can mainly be interpreted as a result of increased frequency of weather systems (circulation patterns) giving rise to high precipitation over large areas. This is in agreement with the conclusion by Caspary [5] who explains the recent very high floods in the upper Danube and upper Rhine by a change in prevailing tracks of weather systems over Northern Europe.

It is indeed very common that floods are high over large territories at the same time. Quasi-periodic phenomena like the sunspot intensity [51] and recently the El Niño Southern Oscillation (ENSO) are some of the possible causes behind fluctuations in the intensities of extreme floods, though the signals are in general rather weak. This varies of course with the geographical location. Krasovskaia et al. [23] studying the ENSO influence on extreme floods in Costa Rica found that also in this case the magnitude was almost constant while the frequency of extreme events was strongly increasing during La Niña years, having in general more frequent tropical storms.

Hydrological design has traditionally been based on the assumption of stationarity. As formulated by Kuusisto et al. [26], virtually all existing water resources systems have been designed on the basis of the same axiom: the past is the key to the future. Facing clear evidence that hydrological time series contain more or less dominant non-stationary features caused by changes/oscillations in climate and large-scale quasi-periodic phenomena, it must be admitted that the worth of data for predictions into the future (design) has decreased [21]. Consequently the uncertainty in such predictions increased and must be assessed in a risk calculation.

The influence of change in land use on hydrological phenomena is also a classical topic in hydrology starting from the studies in small drainage basins in Switzerland at the beginning of this century [6]. Rodda [36] gives an extensive overview. It is clear from such studies that a change in land-use, say deforestation, might have a significant influence on floods. The general conclusion from another famous study in Wagon Wheel Gap [3] that forest reduces the magnitude of ordinary seasonal floods and that forest maintains streamflow in dry weather is a well accepted result, which has been confirmed by many other studies for temperate climates. The critical point is how this finding can be scaled up to large drainage basins. The question of scale is central in a double respect - the scale of the change of land use and the scale of the prevailing weather phenomena. In Scandinavia drastic large scale changes in land-use are fortunately not actual. In other parts of the world the situation is different.

Risk perception, nearly neglected in earlier flood risk studies, has gained more and more attention in recent research. Krasovskaia et al. [24], for example, addressed this topic in another study in the frame of the HYDRA project, trying to get a better insight into the problem and including this important component of flood risk assessment. The study has identified a number of discrepancies in the perception of flood risk by the broad public and decision-makers, which contribute to the total uncertainty in a flood situation increasing thus the total risk.

3. The concept of risk

A generally used definition of risk is the Bayesian expected loss $\rho(a)$ for an action a [4]:

$$\rho(a) = \int L(\varsigma_p, a) f_z^*(\varsigma_p) d\varsigma_p \qquad (1)$$

where $L(\zeta_p,a)$ is a loss function (damage function), ζ_p is the state of nature (e.g. river water level or river discharge with the exceedence probability p) and $f_Z^*(\zeta)$ is the frequency function of ζ_p. In Bayesian terminology this distribution should be the posterior distribution, updated with supplementary information in relation to a prior distribution. Berger also defines a risk function (the frequential risk), which for the present application is formulated as:

$$r(\varsigma_p,\delta) = \int_X L(\varsigma_p,\delta(\mathbf{x})) \, f_Z(\hat{\varsigma}_p(\mathbf{x})|\varsigma_p) \, d\mathbf{x} \qquad (2)$$

where $\delta(\mathbf{x})$ is a decision rule based on the vector of observations $\mathbf{x}$ and $f_Z(\hat{\varsigma}_p(\mathbf{x})|\varsigma_p)$ the sample distribution of the estimate of ζ_p.

The point of departure for calculation of risk of flooding is as a rule a model (regional frequency curve) for the flood discharge $f_Q(q_A)$, where A denotes the area of a drainage basin. This constitutes the prior distribution for our problem. Specifying a certain drainage basin A_i for a site i in a river, the regional model can be updated with local information at this site leading to the posterior distribution. The discharge q_{Ai} can be related to flood stage by the rating curve $q_A = g(\zeta_i)$. From the flood stage the damage (loss) L can be calculated from a stage-damage curve $l = l(\zeta_i)$. Note that all the three variables – discharge, Q, water level, Z and damage loss, L - are stochastic variables. The transformation from the peak discharge to monetary-flood loss is then given by the following equation:

$$l = l[\varsigma] = l[g^{-1}(q)] \, . \qquad (3)$$

Assuming that the peak discharge Q distribution is known, the analytical derivation of the distribution functions for Z and L, given the random variable Q and the transformations l and g^{-1}, is straightforward:

$$F_Q(q) = F_Z(g^{-1}(q)) = F_L(l(g^{-1}(q))) = F_L(l) \, . \qquad (4)$$

The risk is evaluated accordingly as:

$$r = \int_0^\infty l f_L(l)\,dl = \int_0^\infty l(\varsigma) f_z(\varsigma)\,d\varsigma = \int_0^\infty l\big[g^{-1}(q)\big] f_Q(q)\,dq \,. \tag{5}$$

Note that the risk estimate, in this case, only reflects the uncertainty due to the inherent random characteristics of the runoff formation process itself.

Though the expected damage (loss) due to flooding is one of the ways to describe risk of flooding, engineers have generally been reluctant to base decisions solely upon expected damages because the expectations can lump events causing modest damage and events resulting in extraordinary damages [44]. A single expected value does not show trade-offs between high-cost/low probability events and low-cost/high probability events. The hydrological literature offers several alternatives, like the partitioned multiobjective risk method (PMRM) [32]: and the partial expected damage function, i.e. the expected damage from all events greater than a threshold [44]. In the risk analysis literature, risk versus damage magnitude is shown for various events [41,25,33,29,31].

While in some approaches [e.g. 44] the point of departure is a river discharge Q (a stochastic variable), in the case of flooding it is more straightforward to start from the observed critical water level ς_c in a river (i.e. an outcome of a stochastic variable Z). Two problems can be then identified, namely estimation of the loss (damage cost) $\hat{l} = \hat{l}(\varsigma_c)$ caused by this high water level and estimation of the probability $\hat{p} = \hat{p}(\varsigma_c)$ of exceeding this water level.

For establishing a stage-damage curve for a floodplain, i.e. the first problem, the major importance belongs to the following factors:

- *channel characteristics* (river flow profiles $\varsigma_u = h(\varsigma_c)$);

- *floodplain characteristics* (number of economic objects N and their topography $F_\Xi(\xi)$; dikes etc. $\xi_{0:}$)

- *economic loss* (as a function of local water level η, $F_C(c/\eta)$)

- *uncertainties* (in topographic data, economic loss estimates, estimated flood profiles).

For the second problem the principal issues are:

- *exceedance probabilities of floods* (magnitude of floods in the main river $F_Q(q_A)$ and in the headwaters $F_Q(q_{ai})$; synchronicity of floods in headwaters $\rho(q_A, q_{ai})$ and effects of river regulation)

- *channel characteristics* (stage-discharge relation at downstream critical section $q_A = g(\zeta_c)$)

- *uncertainties* (in estimation of exceedance probabilities of floods and in the stage-discharge relation).

Herein an approach for establishing the risk of flooding based on the factors listed above is presented.

4. The stage-damage curve

Resistance to flooding is directly connected to the topographic position of objects (buildings etc), identified by their absolute altitude ξ_i. The distribution of the altitudes (ground floor level of the building) of about 3870 buildings along the 32 km section of the Glomma River has been analysed (Fig. 1, right). The river section along the floodplain was partitioned into 18 clusters of settlements, eight of which have flood protection (dikes). The natural slope of the river along the section is about 0.08 permille at natural water stages and 0.14 permille at high floods (the 1995 flood), which corresponds to a difference in water level of 2.07 m and 3.52 m respectively, between the downstream and upstream boundaries.

The empirical data are used directly for constructing the cumulative distribution of altitude of buildings $F_\Xi(\xi)$ (the distribution of the sample, Cramér, 1971):

$$F_\Xi^*(\xi) = n/N \tag{6}$$

where n denotes the number of sample values ξ_i that are $\leq \xi$ of the total N values.

4.1 CONSEQUENCE VALUE DETERMINATION

A local critical water level for flooding is defined as the depth $\eta_i = \zeta - \xi_i$ for an object (building) i. As long as the water level ζ in the river is low, η_i is negative. When this level is higher than the ground floor of the building, η_i is positive. The exact value of η_i resulting in flooding of an object varies with the type of construction and local topography. The damage by the floodwater is proportional to η_i, the higher the level η_i the higher the damage, until some critical level corresponding to the total damage.

The flooding of houses is the main reason for loss (damage) due to extremely high water levels in a surrounding floodplain with an urban settlement. As evident from Fig. 2, loss value for an individual house is related to the height of the local water level η, though there is a considerable scatter in the data. The augmenting of the costs starts already at negative values of η, i.e. increase in the floodplain groundwater level, raising the likelihood of moisture damages in basements and cellars. The costs shown in Fig. 2 are those paid for flood damage in 1995 by insurance companies, whose agents base the reimbursements only on the depth of water; thus other factors that might have influenced the severity of damage do not show up.

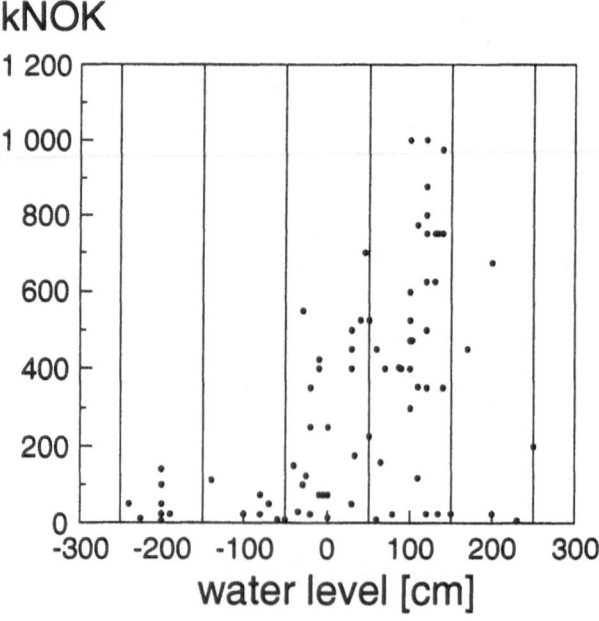

Figure 2. Relation between local water level η and individual cost of damage l' (from [48]).

Due to a large scatter the relation in Fig. 2 is not reasonably well approximated by a deterministic function; and a distribution function conditioned on the height η is suggested instead. The cost for damage L', for an individual house, is in this case looked upon as a stochastic variable with the pdf $f_L(l'/\eta)$. For low values of η (negative) the probability for relatively low costs (with a modal value at zero) is dominating indicating a J-shaped distribution. For increasing values of η the centre of gravity of the costs moves away from the zero axis, giving rise to a clock formed distribution. The distribution is asymmetrical with a tail towards high costs (positive coefficient of skewness). For the time being, due to the large scatter and for simplicity, the cost is taken as expected values $E[l']$ and standard deviation $D[l']$ for only two classes of the local water level (a further subdivision is possible):

- $-2.5 < \eta \leq -0.5$; $E[l'] = 48.1$ kNOR; $D[l'] = 41.7$ kNOR   (low damage)   (7)

- $-0.5 < \eta$; $E[l'] = 411.9$ kNOR; $D[l'] = 293.9$ kNOR   (high damage)

4.2 NUMBER OF FLOODED OBJECTS

When all objects on a floodplain are considered, the levels $\eta_i, i=1,..,N$, where N is the total number of objects (buildings), can be looked upon as a random variable H dependent on the river stage ζ. It is characterised by its conditional cdf (pdf) $F_H(\eta/\zeta)$ $(f_H(\eta/\zeta))$. It is further necessary to consider a variation of the water level along a river section. Analyses have shown no dependence between the altitude of buildings and the south-north direction (the main direction of the river). We can thus assume that the water level in the river and the altitude of buildings are independent. For a river section the water level in the river can be represented by a rectangular distribution:

$$f_z(\zeta) = \frac{1}{\zeta_u - \zeta_d}$$
(8)

where ζ_u and ζ_d are the water levels at the upstream and downstream points, respectively, of the river section. The water levels in the river are all established in relation to the gauging station at Nor (see Fig.1). The following relation has been developed for calculating the water level ζ_L at a distance L metres upstream of Nor:

$$\zeta_L = \zeta_{Nor} + L(1.082\zeta_{Nor} - 151.78)10^{-5} . \tag{9}$$

The water level in a building $\eta = \zeta - \xi$ is determined as the difference between two random variables ζ and ξ. Utilising well known relations from probability theory (see for instance [40]) the frequency distribution of η is formally derived as:

$$f_H(\eta|\zeta_u, \zeta_d) = \frac{1}{\zeta_u - \zeta_d}[F_\Xi(\zeta_u - \eta)]; \qquad \zeta_d - \xi_0 \le \eta < \zeta_u - \xi_0$$

$$f_H(\eta|\zeta_u, \zeta_d) = \frac{1}{\zeta_u - \zeta_d}[F_\Xi(\zeta_u - \eta) - F_\Xi(\zeta_d - \eta)]; \quad \eta < \zeta_d - \xi_0 \tag{10}$$

where ξ_0 is the lower bound of the distribution of altitude of buildings. It thus follows that the derived distribution has an upper bound equal to $\eta_{max} = \zeta_u - \xi_0$, the largest water depth in an individual building. It should be noted that when the distance between the upstream and downstream points is small, the distribution of eq. (10) has the limiting form:

$$f_H(\eta|\zeta_u, \zeta_d) = \frac{1}{\zeta_u - \zeta_d}[F_\Xi(\zeta_u - \eta) - F_\Xi(\zeta_d - \eta)] \longrightarrow f_\Xi(\zeta - \eta)$$

$$for \quad \zeta_u \longrightarrow \zeta_d = \zeta . \tag{11}$$

In the following, the relation of eq. (9) will be utilised and all water depths in buildings will be related to the river stage at Nor, ζ_{Nor}, and the distribution of eq. (10) expressed as $F(\eta/\zeta_{Nor})$. For validation of the statistical approach, it was compared with the corresponding distributions for positive values of η derived from a terrain model in a GIS framework. The agreement was very good (for more details see [10]).

For a fixed critical value of η_c the number of buildings n (of a total N) with a local water level $H > \eta_c$ is evaluated from:

$$n(\eta_c|\zeta_{Nor}) = N \int_{\eta_c}^{\eta_{max}} f_H(\eta|\zeta)d\eta \tag{12}$$

and the number of houses Δn within an interval $\Delta \eta$ around (η_c / ζ_{Nor}) is:

$$\Delta n\left(\eta_c | \zeta_{Nor}\right) = N f_H \left(\eta_c | \zeta_{Nor}\right) \Delta \eta . \tag{13}$$

4.3 EFFECT OF UNCERTAINTY IN TOPOGRAPHY

The topographic data for buildings is evaluated from maps with a contour interval of 1 metre. Maps with contour intervals of 5 and 10 metres have also been used to test the influence of the precision of the map information. The precision of map data σ_M can be estimated to 30% of the contour interval [49]. To be able to evaluate the influence of the precision in the estimation of number of flooded buildings, the empirical distribution eq. (6) is replaced by (it is assumed that the errors in map data follow the normal distribution)

$$F_\Xi(\xi) = \frac{1}{N} \sum_{i=1}^{N} \Phi\left(\frac{\xi - \xi_i}{\sigma_M}\right) \tag{14}$$

where $\Phi(.)$ is the cdf of the normal distribution. Analyses of the errors due to wrong classification of the damage showed that these are around 10-15 % for maps with a 1 m contour interval, 20-30 % for 5 m and 40-50 % for 10 m.

4.4 STAGE-DAMAGE CURVE FOR THE FLOODPLAIN AT NOR

The number n, of a total N, of buildings that will be damaged by flooding on a floodplain increases with the increasing level ζ of the river, as shown by eqs. (12) and (13). The expected number of buildings with local water level $\geq \eta_c$ when the river stage is ζ is $\Delta n(\eta / \zeta_{Nor})$ (eq. 12). The expected damage cost for individual buildings is $l'(\eta)$ (eq. 7), and the expected total cost $l(\eta, \zeta)$ is thus

$$l(\eta, \zeta) = \Delta n\left(\eta | \zeta\right) l'(\eta) . \tag{15}$$

The distribution function of $l(\eta, \zeta)$ is evaluated from eq. (15) as

$$f_L\left(l | \eta, \zeta\right) = f_{L'}\left(l(\eta, \zeta) / \Delta n(\eta, \zeta)\right) / \Delta n(\eta, \zeta) . \tag{16}$$

The expected damage $E[L / \zeta]$ for a certain interval of critical damage, say $\eta > \eta_c$, is derived by integration over the critical interval as

$$E[L|\zeta] = \int_{\eta_c}^{\eta_{max}} l \, f_{L'}(l(\eta,\zeta)/\Delta n(\eta,\zeta))/\Delta n(\eta,\zeta) d\eta . \tag{17}$$

For the present case, when the distribution of individual damage cost is for simplicity described by only two values for high and low damage respectively (eq. 7), the expected damage can be evaluated from

$$E[L|\zeta] = \begin{cases} n(-2 < \eta \le -0.5|\zeta) l'_{low} \\ n(\eta > -0.5|\zeta) l'_{high} \end{cases} . \tag{18}$$

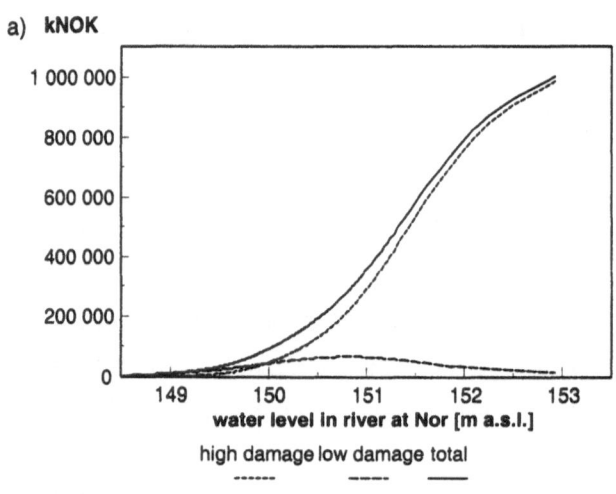

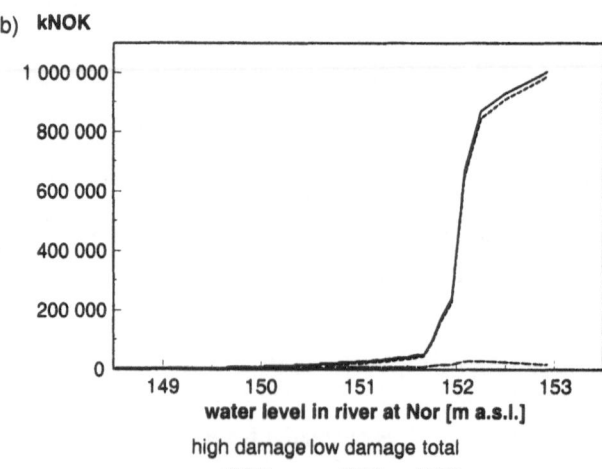

Figure 3. Stage-damage curve for the floodplain upstream of Nor based on 1 m contour interval map data, a) without dikes and b) with account of existing dikes.

This function - the stage-damage curve- a function of the water stage, ζ_{Nor}, at Nor is shown in Fig. 3 for low and high damage and the sum of the two - the total damage (1m contour interval data). It is mainly a reflection of the number of damaged buildings. The uncertainty due to erroneous number of damaged buildings stemming from errors in attributing a damage class, is also relevant for the stage-damage-curve. A further source of uncertainty is added, as revealed by the standard error in the estimates of damage costs for individual buildings. It is large when only a few objects are concerned (e.g. it can be as high as 36% for the high damage class, no dikes) but its relative importance is significantly reduced as the number of damaged buildings increases (e.g. it drops to about 2% for the high damage class, no dikes).

5. Flood stage exceedance probability

The probability of exceeding a critical water level in the main river is:

$$P(Z > \zeta) = 1 - P(Z \leq \zeta) = 1 - F_z(\zeta) . \tag{19}$$

This water level corresponds to a certain discharge q in the river determined by the stage-discharge relation $g( )$ for the river site:

$$q = g(\zeta) \quad or \quad \zeta = g^{-1}(q) . \tag{20}$$

The probability of exceedance of corresponding values of q and ζ are of course identical, i.e.

$$P(Z > \zeta) = P(Q > q), \text{ equivalent to } F_z(\zeta) = F_Q(q) . \tag{21}$$

The relation between the corresponding pdfs is:

$$f_z(\zeta) = f_Q\left(g^{-1}(q)\right) \frac{dg^{-1}(q)}{d\zeta} . \tag{22}$$

5.1 A REGIONAL FREQUENCY CURVE FOR FLOODS IN THE GLOMMA RIVER

Floods and their distribution function $F_Q(q)$ are in Norway determined by so-called regional growth curves (cdfs)[47]. During the last few years a new theoretical framework has evolved aimed at understanding the stochastic structure of regional floods in terms of their physical generating mechanisms. This framework for regional flood frequency is based on contemporary ideas of scaling invariance [39,14].

Here, an approach for constructing a regional flood frequency curve by Gottschalk and Weingartner [13] is applied, which is based on scaling relations of regional floods in terms of L-moment and expected order statistics. These are used to derive expressions for the scale dependence of the parameters of the General Extreme Value (GEV) distribution with the Gumbel (Extreme Value type 1, EV1) as a special case. Three different data sets have been prepared from 35 daily observation records of at least 15 years from the Glomma basin, namely annual maximum series (AMS), partial duration series (PDS) (on the average 3 events per year) and synoptic partial duration series (SPDS) for flood events in 1966, 1967 and 1995. In the latter synoptic data set, the events at the most downstream station have guided the selection of upstream events contributing to the flood downstream within a time lag of up to 8 days. The data cover a range in drainage areas from 40 km^2 to 40000 km^2.

Fig. 4 shows the derived regional frequency curve for PDS, which are based on more events and thus considered to be more reliable. (The regional curve has also been constructed using AMS with only slight differences compared to the PDS). The maximum flood for different probabilities of exceedance and for different scale (area in km^2 divided by 100) is expressed as normalised values with respect to the index flood (the mean flood). The normalised flood is strongly dependent on the scale especially for small probabilities of exceedance. The size of the drainage basin has a dampening effect on the relative magnitude of the flood. A 1:100 year flood can be expected to be 2 times as big as the index flood when the basin is 50000 km^2, but almost 3 times as big for a basin of 10 km^2. For the 1:1000-year flood the corresponding values, with which the index flood should be multiplied, are 3 and 5 respectively.

It is seen in Fig. 4 that the observations in the main branches in the downstream part (for large areas) are close to each other, indicating that the 1966 and 1967 floods had probabilities of exceedance of the order of 1/50. The 1995 flood had a probability of exceedance of around 1/100. The observed floods for the largest area (40221 km^2) indicate higher probabilities of exceedance. A 1:100 year flood observed in a

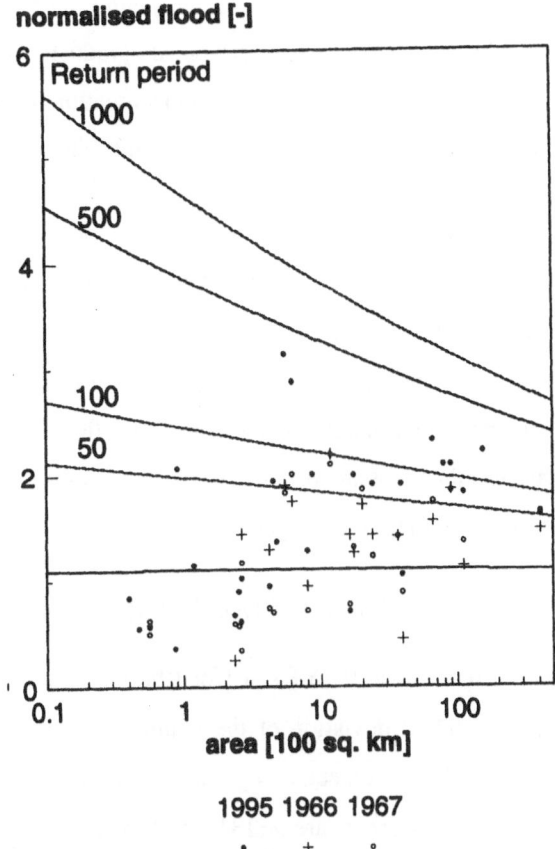

Figure 4. A regional frequency curve for the Glomma river for the normalised flood with respect to the mean flood. The scale factor is expressed as the area in km² divided by the area of a reference basin of 100 km². Empirical data from the floods in 1966, 1967 and 1995 are also shown.

downstream part of a basin does not imply that the floods upstream have the very same probability of exceedance. It is rather on the contrary.

Changes in land use in headwaters are often pointed out as the main reason behind the frequent flooding in Europe during recent years. However, the present study gives evidence that the individual magnitude of floods in headwaters is a factor of a second order to explain downstream flooding. The eventual effects of land use changes in headwaters will be totally hidden by the influence of the primary factor behind downstream extreme floods, namely the synchronicity in time of flood events in headwaters. A flood with, say, a 1/5 probability of exceedance in a headwater is likely to happen every year somewhere in a drainage basin but rather rarely over a large territory at the same time and when this happens an extreme flood situation may occur downstream. The events for downstream gauges in the main river branches are almost identical for PDS and SPDS, but this similarity disappears gradually with a diminishing size of a basin, i.e. the events contributing to the extreme situation in the main rivers do not belong to the population of extreme events for the small basins.

Analysis of the spatial correlation structure of the extreme flood has shown that the more extreme the downstream event the more correlated is it with the upstream events, i.e. the larger is the scale of the causative factors. Thus, the major cause behind the increasing frequency of extreme floods should be rather traced to changes in the phenomena influencing the synchronicity in the timing of floods in headwaters, for example the large scale weather systems bringing rapid temperature change, triggering snowmelt, and/or intense precipitation over vast areas.

5.2 A LOCAL FREQUENCY CURVE FOR FLOODS IN THE GLOMMA RIVER AT NOR

Using a short observation record for 1975-84 at Nor (the maximum in 1995 was also available), the mean and standard deviation of the annual maximum flood have been estimated as 1564.4 and 541.0, respectively. The three parameters in the GEV distribution estimated by the L-moments are u=1313.9, α=333.9 and k=-0.150, with the standard error in these local parameter estimates equal to 39.5, 20.9 and 0.270 ("jackknifing" [27]). The regional model gives an estimated mean value of 1599.7 and the following GEV parameters: u=1405.4, α=337.0 and k=-0.026.

Analysis of the estimation errors showed that for the location parameter u the local estimate has the smallest estimated error. For the scale parameter α the local and regional estimates have almost identical errors. The shape parameter k, finally, is by far best estimated from the regional model, while the local estimate is very uncertain judging by the standard errors. The short local series can help to improve an estimate of centrality and eventually scale, while the form parameter, determining the shape of the tail for extremes, can hardly be helped by the local short series.

Utilising Bayesian principles the regional prior estimates can be updated by the local ones to derive a set of posterior parameter estimates by weighting in accordance with the inverse of the respective error variances [27]. These posterior estimates have been calculated to be u=1318.7, α=335.5 and k=-0.0289 and the corresponding standard errors are 38.0, 14.9 and 0.042. The errors in the estimates for the first two parameters are thus significantly reduced while the error of the third parameter k is not changed at all. The posterior parameter estimates are used in the following to determine the distribution function of discharge at Nor. Applying the transformation eq. (22) to the stage-discharge relation at Nor the distribution of the river stage is derived. These two distribution functions are shown in Fig. 5.

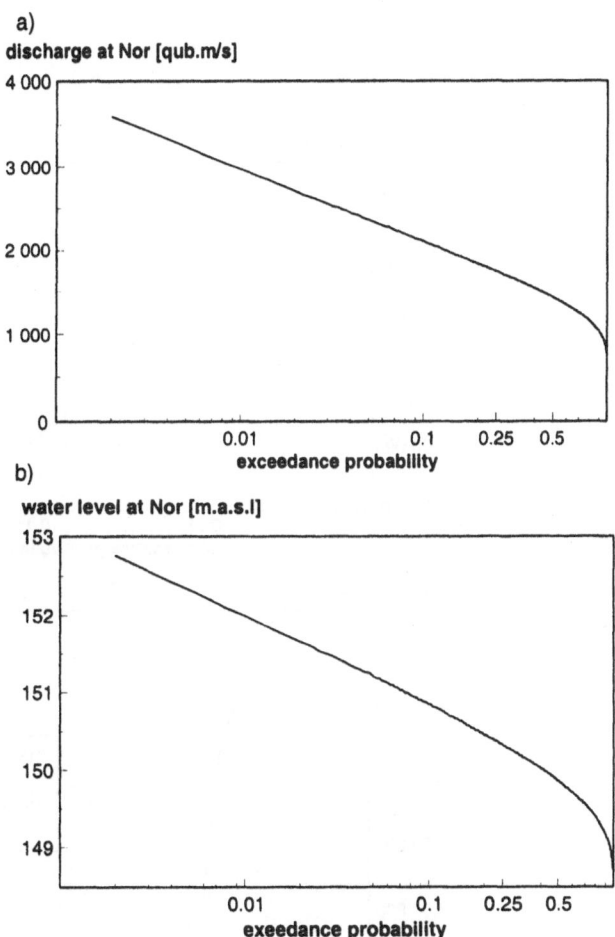

Figure 5. The distribution functions for a) discharge and b) flood stage at Nor.

Though the probability of exceedance (and the return period) is well-established in hydrology, there is a large uncertainty connected to these estimates. Hopefully, the created regional flood frequency curve for the Glomma offers a more consistent tool for estimations for the whole basin, minimising the uncertainties.

6. Risk estimation

Fig. 6 offers an illustration of the concluding results of this study, i.e. the damage distribution and the calculated risk of flooding for the study area. The stage-damage curve (Fig.3) and the distribution curve of river stage (Fig. 5b) are thus the background for constructing the damage distribution. The risk estimate is derived as the expected value of the damage distribution

$$r = \int_0^\infty lf(l)dl \qquad (23)$$

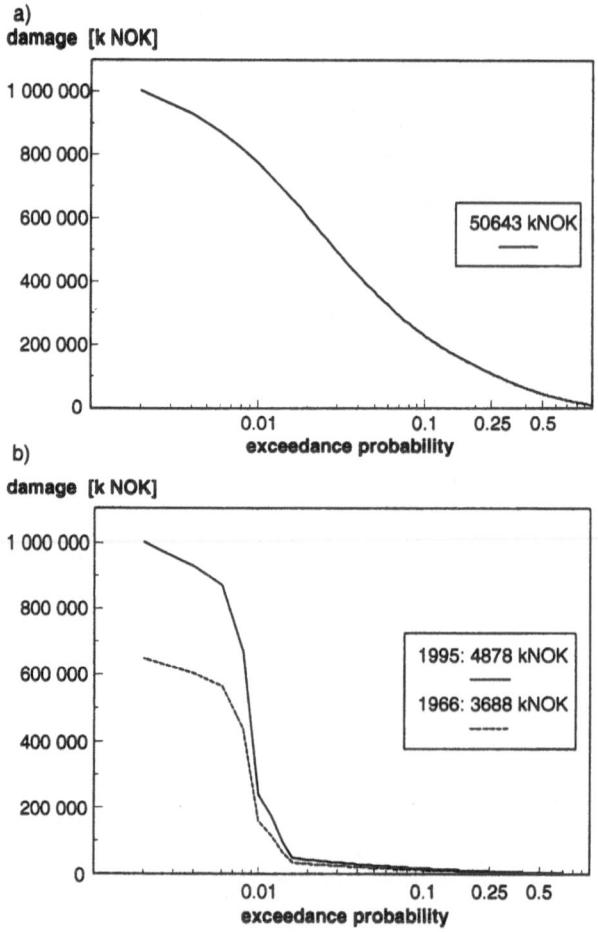

Figure 6. The distribution of damages for the floodplain upstream of Nor. The full line in a) shows the case for the assumption of the absence of physical obstacles preventing the river water from reaching a building at a certain altitude. The full line curve in b) is for the present (1995) situation when the height of the dikes constructed after the 1967 flood has been accounted for. The dashed line shows a hypothetical curve for the number of buildings that existed in 1966.

Fig. 6 presents three cases. The first one, illustrated in Fig.6a, shows the damage distribution for a hypothetical situation with no dikes. In such a case the curve is declining smoothly from high damages for low exceedance probabilities to moderate and low ones for high exceedance probabilities. The risk is estimated to be 50 million Norwegian crowns (NOK). This value might seem low in relation to the very high damages of up to 1 billion NOK for the exceedance probability of 1/1000. This reflects actually the criticism expressed by Thompson et al. [44] and others, who noted that the risk estimates give too much weight to very frequent but low damage causing events.

The two curves in Fig.6b consider the situation with the existing dikes. The curve for 1995 offers the damage description relevant for the year of the flooding. The curve for 1966 is a hypothetical one, where the number of buildings in the area has been reduced to their actual number that year. At that time the dikes were less extensive than in 1995 but this is not accounted for in the figure. The respective risk for the two cases is 4.9 million NOK and 3.7 million NOK. These figures are very low compared to the situation shown in Fig.6a. This is obviously due to the fact that the damage is very low below the critical levels of the dikes, which, as it can be seen, represents approximately the 1:100 year flood. The damage for the flooding with lower exceedance probability is as high as in Fig.6a for the 1995-curve.

7. Discussion and conclusions

A preliminary unofficial estimate of the actual damage costs of the 1995 year flood for the Åsnes and Grue municipalities is 130 million NOK (Hallvard Berg NVE, personal communication), 80 million NOK of which concerns damage to buildings. Identifying this latter sum on the damage curve for 1995 in Fig. 6b, it is seen that it corresponds to a break point on the curve, where it starts to increase sharply at an exceedance probability of 1.5 % (i.e. a return period of 67 years). This would indicate that some dikes have been overtopped and areas protected by them were flooded. This interpretation is in agreement with what really happened. During the flooding the dikes were reinforced and topped with sand bags to withstand a higher river stage. The hypothetical damage that would have occurred without any reinforcements is probably significantly higher than the actual damage closer to the 1 % exceedance probability.

It is a difficult task to test and validate in a formal way the damage distribution curves in Fig. 6. We can only conclude from this very rough comparison that the 1995 curve is in agreement with what really happened and that the values of damages are in reasonable agreement with regard to the uncertainties in the estimate of the damage distribution curve. Some of these uncertainties are inherent to the problem and have to be accepted. Nature behaves in an unexpected way and surprises are common.

Uncertainties due to limited amount of data are always a critical point. Economic constraints impose a continued use of existing hydrologic and climatic observation networks and thereby this part of the uncertainty cannot be reduced either. For other types of data (topographical, economic etc.) a development towards a better resolution and larger samples can be expected. There are also several ways of improving the methodology presented here to obtain more reliable results.

Among the many factors that influence the risk of flooding, only those classified as floodplain characteristics - economic activity, topography of objects (buildings), flood protection works, economic loss - play a significant role. This is clearly illustrated by the three curves in Fig. 6. The diagrams show how the risk changes with increased economic activity (compare the curves for the 1966 and 1995 year floods) and with the construction of dikes (compare the curve in a) with those in b)). Other factors like land use changes in upstream headwaters and non-stationarities in flood records have in comparison only a secondary influence in the risk estimation. Their contribution is difficult to separate from the different sources of statistical uncertainty.

The available data on damage costs related to water stage in flooded objects are quite unique for Norway and, though the sample size is rather small, they are very valuable for this study. However, it would be desirable to test the method on more extensive data. Though the analysis focused on damage to buildings, the principles developed can be applied also to other types of object linked to topography like roads, railways, agricultural land etc. The number of damaged buildings and damage per building, respectively, need then to be replaced by, for example, length of damaged roads and damage cost per length unit.

The resolution in topographic data is of a vital importance for reliable estimates of risk of flooding. It is clearly shown here that such data taken from maps with 5 and 10 m contour intervals are not good enough and results are highly unreliable. Topographic data from maps with 1 metre contour intervals are acceptable but the uncertainty is still significant.

The flood line calculations have been rather superfluously treated in this study and a very simplified model has been used. Such calculations have been performed by other research groups in the HYDRA research programme. However, the results have not been available in time to be utilised here.

The methodology developed for establishing a regional frequency curve and for updating such a model with local information is probably as far as one can reach at the present state-of-the-art to minimise the uncertainty in flood estimation at a specific site. The total uncertainty is still substantial and the part related to the choice of model needs further clarification. Only adding more observation points in space can, however, significantly reduce the total uncertainty. The local updating procedure calls for the establishment of new observation sites at places where the present precision is considered unacceptable. The derived distribution function approach may offer better alternatives for the future, while the use of the present generation of conceptual models cannot do this. The uncertainty is 40%-100% higher when such models are used for flood estimation compared to a regional frequency curve as reported by Watt and Paine [49].

The damage distribution curves have been established for a rather limited area. For a larger area the task is naturally more complex. A damage distribution curve can be calculated with the methodology developed for each area with a potentiality for being flooded. The difficulties arise from the need for a correct account of the synchronicity of flood events over large areas - how large is the area where we can expect a 100-year flood at the same time? The problem is related to another one, namely how information on upstream events can be utilised for improving the precision in estimates of extreme floods downstream. For the Glomma River this question has special relevance as two equally large branches join together into one river reach. The analysis of the synchronicity of floods briefly presented here is a first modest step to clarify these questions.

8. References

1. Ashkar, F. and Rouselle, J. (1981) Design discharge as a random variable: a risk study. Water Resources Research 17(3): 577-591.
2. Aven, T. (1992) Reliability and risk analysis. Elsevier Applied Sciences.
3. Bates, C.G. and Henry A.J. (1928) Forest and stream-flow experiment at Wagon Wheel Gap, Colorado. Monthly Weather Review supplement No 30 74pp.
4. Berger, J.O. (1985) Statistical decision theory and Bayesian analysis. Second edition, Springer.
5. Caspary, H.J. (1998) Regional increase of winter floods in southwest Germany caused by atmospheric circulation changes (abstract). Annales Geophysicae Part II Hydrology, Oceans & Atmosphere, Supplement II to Volume 16 p. C463.
6. Engler, A. (1919) Einfluss des Waldes auf den Stand der Gewasser. Mitt. Schweiz. anst. für das Forsliche Versuchswesen 12, 626.
7. Galea, G. and Prudhomme, C. (1994) Notions de base et concepts utiles pour la compréhension et la modélisation synthétigues des régimes de crue des bassins versants au sens des modèles QdF. Revue des sciences de l'eau 10(1):83-101.
8. Gilard, O. (1998) Les bases techniques de la méthod Inondabilité. Cemagref Édition.
9. Gottschalk, L. and Krasovskaia, I. (1997) Climate change and river runoff in Scandinavia, approaches and challenges. Boreal Environment Research, 2(2): 145-162.
10. Gottschalk,L. and Krasovskaia,I. (2000) Economic risk of flooding. A case study for the floodplain upstream Nor in the Glomma river, Norway. HYDRA-report no R01, Norwegian Water Resources and Energy Directorate, in press.
11. Gottschalk,L.and Sælthun,N.R. (1990) Large scale temporal variability and risk of design failure. (Abstract) Annales Geophysicae, EGS XV General Assembly, Copenhagen, 1990.
12. Gottschalk, L. and Weingartner, R. (1998) Distribution of peak flow derived from a distribution of rainfall volumes and runoff coefficient, and a unit hydrograph. Journal of Hydrology 208:148-162.
13. Gottschalk, L. and Weingartner, R. 1998: Scaling of regional floods - an L-moment approach. European Geophysical Society XXIII General Assembly Nice, 20-24 April. Abstract. European Geophysical Society, Annales Geophysical, Part II, Hydrology, Oceans & Atmosphere, Supplement II to Vol. 16.
14. Gupta, V.K., Mesa, O. and Dawdy, D.R. (1994) Multiplicative cascades and spatial variability of rainfall, river networks and floods. In: Rundle, J., Klein, W. and Turcotte, D. (Eds.) Reduction and predictability of Natural Disasters. SantaFe Institute, XXV, Addison-Wesley.
15. Haan, C.T. (1977) Statistical methods in hydrology. The Iowa State University Press, Ames, Iowa 378 pp.
16. Handemer, J.H. (1983): Risk information for floodplain management, Discussion. Journal of Water Resources Planning and Management ASCE, 114:120-123.
17. Jain, S.K., Yogararasin, G.N. and Seth, S.M. (1992) A risk-based approach for flood control operation of a multipurpose reservoir. Water Resources Bulletin 28(6): 1037-1043.
18. James, L.D. and Hall, B. (1986): Risk information for floodplain management. Journal of Water Resources Planning and Management ASCE, 112:485-499.
19. Jaeggi, M.N.R. and Zarn, B. (1990): A new policy in designing flood protection schemes as a consequence of the 1987 floods in the Swiss Alps. In W.R. White (ed.) International Conference of River Flood Hydraulics, Hydraulic Research Limited, John Wiley & Sons Ltd. pp 75-84.
20. Kite, G.W. (1977) Frequency and risk analysis in Hydrology. Water Resources Publications, Fort Collins, Colorado.
21. Krasovskaia, I. (1996) Stability of river flow regimes. Dr. Philos. thesis, Department of Geography, University of Oslo, Norway, Rapport No 5, Oslo.
22. Krasovskaia,I. & Sælthun,N.R. (1997) Sensitivity of the stability of Scandinavian river flow regimes to a predicted temperature rise. Hydr.Sci.J. 42 (5), 693-711
23. Krasovskaia, I., Gottschalk, L., Rodrigues, A. and Laporte, (1999) Dependence of the frequency and magnitude of extreme floods in Costa Rica on SOI. IAHS Publication. 255, 81-89.
24. Krasovskaia,I., Gottschalk,L., Saelthun,N.R, & Berg,H. (2000) The HYDRA R4 project: improved front line decision tools - risk perception among decision-makers and riparian population (in Norwegian). Report, Norwegian Water Resources and Energy Directorate. (in press).
25. Kundzewicz, Z.W. (1998) Floods of the 1990s: Business as usual? WMO Bulletin 47(2):155-160.
26. Kuusisto, E., Lemmelä, R., Liebscher, H. and Nobilis, F. (1994) Climate and water in Europe: some recent issues. WMO RAN Working group on hydrology, Helsinki.
27. Kuzera, G. (1982) Combining site specific and regional information: An empirical Bayes approach. Water Resources Research 18(2):306-314.

28. Lambert, J.H. and Li, D. (1994) Evaluating risk of extreme events for univariate loss functions. Journal of Water Resources Planning and Management ASCE 120(3): 382-399.

29. Loucks, D.P., Stedinger, J.R. and Haith, D.A. (1981) Water resources systems planning and analysis. Prentice-Hall, Inc. N.J.

30. Lowrance, W.W. (1976) Of acceptable risk. William Kaufmann inc., Los Altos, California.

31. Mays, L.W. and Tung, Y.-K. (1992) Hydrosystems engineering and management. McGraw-Hill, Inc. New York.

32. Mitsiopolous, J., Haimes, Y.Y. and Li, D. (1991) Approximating catastrophic risk through statistics of extremes. Water Resources Research 27(6). 1223-1230.

33. Parett, N.F. (1987) Bureau of Reclamation use of risk analysis. In Singh, V.P. (ed.) Application of frequency and risk in water resources. D. Reidel Publishing Co. Boston, Mass. 411-428.

34. Plate, E. J. (1998) Probabilistic design for flood protection structures. In Casale, R., Rasmussen, N.C. (1981) The application of probabilistic assessment technology to energy technologies. Annual Rev. of Energy 6:123-138.

35. Rasmussen, P.F. (1991) On the sampling distribution of exceedance probabilities. In Ganulis, J. Water resources engineering risk assessment. NATO ASI Series Vol. G29: 91-96.

36. Rodda, J.C. (1976) Facets of Hydrology. John Wiley & Sons.

37. Rodda, J.C. (1995): Wither World Water. Water Resources Bulletin 31(1):1-7.

38. Simonovic, S.P. and Mariño, M.A. (1981) Reliability programing in reservoir management 2. Risk loss functions. Water Resources Research 17(4): 822-826.

39. Smith, J. (1992) Representation of basin scale in flood peak distributions. Water Resources Research 28(11):2993-2999.

40. Springer, M.D. (1979): The algebra of random variables. John Wiley & Sons.

41. Starr, C. (1969) Social benefit versus technological risk. Science 165 (3899): 1232-1238.

42. Stedinger, J.R. (1997) Expected probability and annual damage estimation. Journal of Water Resources Planning and Management ASCE 123(2): 125-135.

43. Tang, W.H., Mays, L.W. and Yen, B.C. (1975) Optimal risk-based design of storm sewer networks. Journal of Environmental Engineering ASCE 103(3): 381-398.

44. Thompson, K.D., Stedinger, J.R. and Heath, D.C. (1997) Evaluation and presentation of dam failure and flood risk. Journal of Water Resources Planning and Management ASCE 123(4): 216-227

45. Tung, Y.-K. and Mays, L.W. (1981) Optimal risk-based design of flood levee systems. Water Resources Research 17(4): 843-852.

46. Tukey, J.W. (1958) Bias and confidence in not-quite large samples. (Abstract) Ann. Math. Stat. 29:614.

47. Tveito, O.E. (1993) A regional flood frequency analysis of Norwegian catchments. Dep. of Geophysics, University of Oslo, Report No 86.

48. Wathne, M. (1998) Samfunnskostnader på grunn av flomskade i vassdrag. SINTEF Rapport, Trondheim.

49. Watt, W.E. and Paine, J.D. (1992): Flood risk mapping in Canada: 1. Uncertainty considerations. Canadian Water Resources Journal 17:129-138.

50. WMO (1995): The global climate system. Climate system monitoring June 1991-November 1993. WMO-No 819, ISBN 92-63-10819-6.

51. Yongquan, W.(1993) Solar activity and maximum floods in the world. In: Kundzewicz, Z.W., Rosbjerg, D., Simonovic, S.P. & Takeushi, K. (eds.) Extreme Hydrological Events: Precipitation, Floods and Drought. IAHS Publ. 213:121-127.